Zeil/Kienberger (Eds.)

Geoinformation for Development

Bridging the divide through partnerships

Geoinformation for Development

Bridging the divide through partnerships

Zeil / Kienberger (Eds.)

UNIVERSITY SCIENCE PRESS

(An Imprint of Laxmi Publications Pvt. Ltd.)

An ISO 9001:2008 Company

BENGALURU • CHENNAI • COCHIN • GUWAHATI • HYDERABAD
JALANDHAR • KOLKATA • LUCKNOW • MUMBAI • RANCHI • NEW DELHI
INDIA • USA • GHANA • KENYA

GEOINFORMATION FOR DEVELOPMENT

Originally published 2005 in German language under the title "Elektronische Sicherheitssysteme. Hardwarekonzepte, Model und Berechnung. " by Prof. Dr. Josef Börcsök. Ist, edition 2005. Herbert Wichmann Verlag, Verlagsgruppe Hüthig Jehle Reh GmbH, Heidelberg, Germany. www.huethig-jehle-rehm.de.
English Edition : "Electronic Safety Systems. Hardware Concepts, Models and Clculations " by Prof. Dr. Josef Börcsök. Ist, editi 2007. Herbert Wichmann Verlag, Verlagsgruppe Hüthig Jehle Rehm GmbH, Heidelberg, Germany. www.huethig-jehle-rehm.de
First Indian Edition 2009

This special low-priced edition is only for sale in India, Nepal, Sri Lanka, Bangladesh and Pakistan.

Printed and bound in India
Typeset at : Kalyani Computer Services, New Delhi
First Edition : 2009, *Reprint* : 2015
UGD-9374-250-GEOINFORMATION FOR DEVELOP
ISBN : 978-81-318-0718-7
Price: ₹ 250.00

Published in India by

UNIVERSITY SCIENCE PRESS
(An Imprint of Laxmi Publications Pvt. Ltd.)
An ISO 9001:2008 Company
113, GOLDEN HOUSE, DARYAGANJ, NEW DELHI-110002, INDIA
Telephone : 91-11-4353 2500, 4353 2501
Fax : 91-11-2325 2572, 4353 2528
www.laxmipublications.com info@laxmipublications.com

Branches

Bengaluru	080-26 75 69 30
Chennai	044-24 34 47 26
Cochin	0484-237 70 04, 405 13 03
Guwahati	0361-254 36 69, 251 38 81
Hyderabad	040-27 55 53 83, 27 55 53 9
Jalandhar	0181-222 12 72
Kolkata	033-22 27 43 84
Lucknow	0522-220 99 16
Mumbai	022-24 91 54 15, 24 92 78 6
Ranchi	0651-220 44 64

C—9331/014/10
Printed at : Print 'O' Pack, Delhi.

Preface

In an open world, political, economic and social challenges are increasingly interlinked. Helping our partners in need and investing in their long-term development is therefore not just a moral responsibility but a political imperative. Europe must export stability and good governance in order to avoid importing instability.

Today, the European Union and its Member States are by far the largest donor of economic and development assistance in the world, providing more than half of global assistance and two-thirds of grant-aid. To ensure that our help is both rapid and effective, up-to-date information on the situation in our partner countries is essential. Without accurate spatial intelligence, reconstructing communities after natural disasters, promoting sustainable development and safeguarding human security is close to impossible.

Therefore, this book is very timely. It draws on the substantial expertise of specialists in the field and compiles conference papers and recommendations from a special session of the annual AGIT symposium held in Salzburg, Austria, on July 7 2006. These analyses cover a broad range of important topics, from poverty reduction to demining, from monitoring food security in crisis regions to the strategic water challenge in the Middle East. They thus provide food for thought for foreign policy practitioners and the wider development community alike.

This book delivers "geo-information for development". As Europe's External Relations Commissioner responsible for the bulk of EC assistance, I am particularly appreciative of that.

Supporting our partners' development is essential to make globalization work. This volume is a well-targeted contribution to this effort.

Dr. Benita Ferrero-Waldner
Commissioner for External Relations and
European Neighbourhood Policy

Introduction

During the final panel discussion of this conference, a panellist defined development *'as the process to achieve and secure peace, prosperity and dignity'*. Taking this further, we have to recognise that without stability there will be no development. Evidence for this notion can be found all around us: Colombia, Darfur, Palestine, Iraq, and not so long ago in Kosovo. Another threat for development to progress is only recently moving to the centre of the debate – the link between ecosystem integrity and human security. Past and present global environmental change, environmental degradation and environmental conflicts affect the lives of millions of people across the world every year. This interplay results in significant lost opportunities in terms of human and economic development, and imperils the process of building democracy, establishing peaceful international relations, as well as reducing lawlessness and human insecurity. At a fundamental level, environmental insecurity originates from environmental degradation, natural hazards and weak coping (adaptive) capacities of individuals and societies exposed to these threats. Klaus Töpfer, the former director of UNEP, captures the interdependencies in his statement: *'Sustainable development is a security imperative. Improving degraded environments and achieving sustainable development enhances human security, prevents conflict, and builds peace'*. At the European Conference on Integrating Environment, Development and Conflict Prevention, held in Berlin in March 2007[1], a policy actor stated that *'the choice of mankind will be either to invest in sustainability or risk violent conflicts on a large scale'*. He further recommends that *'industrialised countries should critically reflect on existing systems and their definition of development and growth before forcing developing countries to adopt these systems'*. The growing awareness that impacts such as of climate change are a global challenge (although, some regions or communities are affected more than others), also offers, at least in an ideal world, great opportunities for cooperation.

Over the last decade, the wordings have changed. Today *'aid'* refers to the support during emergencies and the provision of humanitarian action in crisis situations. When referring to development processes, the term *'assistance'* reflects a change of perception. Aid is being given, often without involvement either physically or mentally. Assistance might indicate cooperation in processes or systems to improve coping capacities and, to be successful, requires mutual learning, understanding and commitment. The sub-theme of this conference expresses what we, the organisers, mean by cooperation: *Bridging the divide through partnerships*.

Partnerships have to be built and only last by ascertaining confidence and mutual trust. A balanced partnership requires matching capacities on both sides, concerning people and institutions alike. Taking mitigation measures in the face of climate change as an example, technology alone will not be sufficient. The best methodologies and tools are ineffective without the comprehension of context, e.g. local ecosystems, traditional management schemes, or cultural values. Capacity building includes technology transfer and sharing of expertise and is the core objective of cooperation projects. The concept of partnerships has

---

1 Integrating Environment, Development and Conflict Prevention. European Conference hosted by the German EU Council Presidency 2007, Berlin March 29 and 30, 2007. http://www.adelphi-consult.com/ECC2007/documents.htm.

now even penetrated policy documents such as the EU Strategy for Africa[2]. It states that *'one important component of the broad-based dialogue is the launching of twinning partnerships bringing together African and European universities and schools, parliaments, towns and cities, municipalities, businesses and industries, trade unions, civil society networks and museums. Another innovative proposal is the creation of a European programme for people with skills to share who want to learn more about, and get involved in, Africa's* (and Europe's, add. by authors) *development'.*

The special session of the annual AGIT symposium (http://www.agit.at) held in Salzburg, Austria, on July 7 2006 was organised under the theme 'Geoinformation for Development (gi4dev)'. Recognizing that accurate geospatial information (GI) is an essential pre-requisite for the achievement of development objectives – the Millennium Development Goals are but one example – the initiative for gi4dev was advocated by the participants of a meeting at the AGIT 2005, where low-cost and free-ware solutions for projects in developing countries were discussed. Methodologies and tools are increasingly made available by researchers and technicians in developed countries, which could support the efficient management of spatial information as well as facilitate the access to this information by civil society in the developing world. Gi4dev seeks to explore opportunities and best practices for technology transfer and development-oriented applications.

'We've missed more environmental crises than we've anticipated. To build peace around these issues, we must monitor and assess environmental data in a timely manner'[3]. The capability to monitoring the changes around us and the skills to handle information required for the evaluation of intervention options are essential conditions for effective decision-making. In this context, the special session gi4dev and this book intend to open a dialogue between providers and users of geoinformation systems with the aim to build partnerships, which can help to bridge divides in technology, knowledge and perceptions.

The book offers an account of the conference in July 2006. The evaluation of the event also identified challenges to be addressed in future: many partners from developing countries were unable to participate due to funding problems or delays in obtaining visa; donor agencies, even though interested, did not take part in the meeting. Therefore, a web-platform 'www.gi4dev.org' is launched with the aim to increase advocacy and, following suggestions made at the conference, the next meeting will be organised at a location in the developing world. We warmly welcome your feedback and suggestions.

We would like to acknowledge the support received from the members of the gi4dev scientific committee, Sharon Gomez, Dusan Sakulski, David Stevens, Hans-Peter Thamm, Tsehaie Woldai, S. Shahnawaz, for evaluating abstracts and designing the programme. Annelene Kammer and Florian Albrecht kept the organisation of the event afloat. At Herbert Wichmann Verlag, we thank G. Olbrich for his patience in waiting for the manuscript.

---

2 EU Strategy for Africa: Towards a Euro-African pact to accelerate Africa's development. COM(2005) 489 final.
http://ec.europa.eu/development/ICenter/Pdf/04_eu_strategy_for_africa_12_10_2005_en.pdf.

3 Environmental Sustainability: Perspectives on the World. Marc Levy, Center for International Earth Science Information Network (CIESIN).
http://cero.columbia.edu/full/marc_levy.html.

Gi4dev was supported by United Nations University, UN Office for Outer Space Affairs, African Association of Remote Sensing of the Environment, and Deutsche Gesellschaft für Technische Zusammenarbeit GTZ.

Salzburg, June 2007 *Peter Zeil and Stefan Kienberger*

Table of Contents

Chapter 1

The Pressing Issues

At the centre of all our efforts to monitor the status of the environment, our livelihoods and our resources, are the data which capture the initial situation (baseline) against which we measure the rate of change. The challenges for many countries are to establish this baseline, to maintain and service databases, to provide access, to facilitate the sharing of this asset, and to develop and strengthen the capacity to extract the essential information required for decision-making. Climate change impacts, socio-economic conditions, land degradation are all forceful drivers influencing the livelihoods and their stability. The articles, which follow, describe specific cases and demonstrate the role geoinformation can play in the process to achieve improvements.

Mozambique: Land cover and land use data at national and provincial levels compose the essential baseline data set for establishing forest inventories. The size of the country and limited financial as well as human resources call for an approach using remote sensing supported by GIS to allow the inclusion of existing knowledge.

Colombia: Expanding agricultural activities threaten forest stands while seldom neither the best crop yields nor marketable qualities are achieved. To raise the income for coffee growers, experts tested a quality label indicating the origin of the product throughout the supply chain and a geo-tracking system linking producers with consumers. By describing spatial and temporal variation of the environment, Geographic Information Systems can assist to detect suitable crops for specific environmental niches and support farmers to find optimal management practices for their plot of land.

Dead Sea Basin: Spatial information on land use and land cover as input to dynamic models facilitate the calculation of development scenario maps of water and land management for the next 20 years – important aspects to strengthen the participation of all actors in the development process with the aim to foster stability in the region.

East Africa: Stable livelihoods we can only maintain if the integrity of ecosystems remains guaranteed; for monitoring biodiversity, as formidable task, local and national expertise has to be built preferably in the framework of collaborative research.

The discussion further focused on the issue of data and data access. Even though data are abundant, the information about their repositories often is incomplete and the hesitation to share widespread. Therefore, actors in cooperation projects should ascertain that data and information are made accessible, either through local and regional organisations, or provided by means of dissemination platforms (e.g. GEONETCast[1]). The sustainability of information systems and their impact on development can only be achieved when the users participate in design and implementation. To this end, projects should strengthen the capacity of local organisations to evaluate implemented systems and modify them to

---

[1] GEONETCast is a near real time, global network of satellite-based data dissemination systems designed to distribute space-based, air-borne and in situ data, metadata and products to diverse communities. http:// earthobservations.org/progress/GEONETCast.

respond to specific needs. Bridging the divide requires confident and knowledgeable institutions on both sides. The partnerships between these institutions represent the means of verification for sustainability, of projects and solutions.

Spatially Explicit Land-Cover/Use Data as Input to an Integrated Assessment of Mozambican Forests

Ferdinando URBANO and Louisa J. M. JANSEN

Summary

Land-cover/use and its changes are important driving forces of (global) environmental change and therefore central to sustainable development. The inventory of the land-cover/use types in Mozambique, their location, extent, distribution and changes at the national level and at a more detailed level in two Provinces, using up-to-date technologies and international standards, provides the government with spatially explicit data and information for the monitoring of natural resources. In order to realize this goal, a conceptual model has been conceived and implemented as a geodatabase that was built on and further extends the local capacities thereby overcoming any missing institutional provisions or inadequate human resources. The land-cover/use data assessment was realized through on-screen interpretation of remotely sensed satellite images, supported by a large set of environmental data and information collected in the field. The work was executed in the context of the Integrated Assessment of Mozambican Forests (AIFM) project for the Mozambican Ministry of Agriculture and Rural Development – Direcção Nacional de Florestas e Fauna Bravia-Unidade de Inventario Florestal.

1 Introduction

1.1 Background

Mozambique, located on the south-eastern coast of Africa, suffered a long civil war (1977–1992). Recurrent droughts in the hinterlands resulted in increased migration of the population to urban and coastal areas with adverse environmental consequences. In 1987, the government embarked on a series of macro-economic reforms designed to stabilise the economy. These steps, combined with donor assistance and with political stability since the multi-party elections in 1994, have led to dramatic improvements in the country's growth rate. In spite of these gains, most of the population in Mozambique remains below the poverty line and subsistence agriculture continues to employ the vast majority of the country's workforce. In its present policies that emphasize the market-oriented economy, the role of the private sector and decentralisation are reflected in the forest policies and legislation. This offers challenging opportunities for the development of sustainable management and underlines the importance of the land-cover assessment in order to monitor its changes and implement subsequent management policies. In the turmoil of a changing economy, the spatial and temporal dynamics of land-cover are continuously evolving. It is, therefore, important for the Mozambican Government to have accurate and timely information for natural resources management, land-use planning and land-use policy development, as a prerequisite for monitoring and modelling land-use and environmental impact.

1.2 The Integrated Assessment of Mozambican Forests (AIFM) project

The Ministero da Agricultura (MINAG) – Direcção Nacional de Florestas e Fauna Bravia (DNFFB) decided to assess the national forest resources and to set up an inventory system capable of providing the forest administration and concerned institutions with an efficient tool for planning forestry sector development and for setting up a strategy for a sustainable use of forest and wildlife resources. In order to create an updated efficient information system for planning and managing land resources, the Government of Mozambique promoted the Integrated Assessment of Mozambican Forests (AIFM) project. The project has been financed by the Italian Foreign Ministry. It has been realized since 2005 in full synergy between MINAG – DNFFB and the executing company.

The general objective of the Integrated Assessment of Mozambican Forests (AIFM) project is the assessment and monitoring of the extent, state and changes of Mozambican forests and wooded lands in a timely and accurate manner. It is organized on two integrated geographic levels of detail: national and provincial.

At the national level the main AIFM's specific objectives are: assessment of the current status and the predictable future development of forest and woodland resources; explanation of the kind, type and degree of changes which have occurred in land-cover and forest ecosystems in different areas; the foreseen consequences of these changes for the human and wildlife conditions, and the measures to be taken to mitigate negative impact of current forest use and improve the living conditions of the local population; description of the main environmental components affecting the current state of forest resources and their interactions, in order to address use conflicts and regulate future development of the forestry and wildlife sector; improvement of the DNFFB-Unidade de Inventario Florestal (UIF) technical and operational capabilities to carry out and supervise forest resources assessment and monitoring activities; to plan new forest inventories and management plans; to analyze and provide timely and reliable information on the status and changes of national forest resources. At the provincial level the AIFM's specific objective is the assessment of the Manica and Maputo provincial Forest Resources at a much more detailed level.

In all these aspects, land-cover/use plays a central role.

1.3 Land-cover/use assessment

Land-cover is defined in the FAO/UNEP LCCS as the "(bio) physical features one can observe on the surface of the Earth" (Di Gregorio and Jansen, 2000). Land-use is defined as "the type of human activity taking place at or near the surface" (Cihlar and Jansen, 2001). The AIFM Project has adopted and applied these two definitions.

Land-cover can be a cause, constraint or consequence of a certain land-use. Land-use is determined by natural, economic, institutional, cultural and legal factors. In general, possible land-uses are limited by (bio) physical constraints. Agricultural practices differ from one region to another and different types of land-use are practised on the same type of land in different areas, depending on the history, local traditions and way of life, apart from the (bio) physical constraints (Cihlar and Jansen, 2001). The location of an area with respect to other land-use types, such as residential and industrial areas, is also an important factor. Economic incentives as part of policy can affect land-use and land-cover patterns.

Within the AIFM project framework, the land-cover/use assessment and change study has a central role. An international team of 10 multidisciplinary (photo-interpretation, land-cover, forestry, GIS, remote sensing) experts has been set up to perform remote sensing land-cover/use based interpretation and analysis tasks. Even though this component was completely integrated within the general framework of the project, it was executed as an autonomous component.

The remote sensing and land-cover component contributes to the overall goals by developing some specific tools, methodologies and data collection techniques (JANSEN et al., 2006). Its specific objectives is the presentation of an overview of occurring land-cover types in Mozambique (both at national and, for Manica and Maputo, Province level), their location, extent and distribution according to a specific legend defined by MINAG – DNFFB, using up-to-date technologies and international standards in land-cover data collection. *De facto*, it provides the baseline data set on which the stratification of the Forest Inventory is based.

Moreover, a pilot study of land-cover change was executed in Manica Province because knowledge about land-cover change has become increasingly important in order to analyse environmental processes and problems. Processes such as deforestation, encroachment of agriculture in forested areas, loss of prime agricultural land and urbanisation etc., must be understood if living conditions and standards are to be improved.

As AIFM is an operation project, the methodology used has been strongly dependant not only on scientific goals, but also on project requirements and constraints.

The main requirements and constraints considered were: the integration of multi level land-cover/use assessment and its application for interpretation and data management/analysis phases; the possible reuse of the methodology by DNFFB – UIF for updates and further developments, ensuring the technical sustainability of the project; the land-cover/use classes (legend) were defined *a priori* and interpretation methods had to be able to detect them; the effective data availability and quality were limited and any possible free data sources connected to land-cover/use had to be considered in order to improve information on land; time limits, combined with the requirement of a national coverage, forced a pragmatic and timely approach.

In order to successfully implement the land-cover/use component, an environmental model framework was developed. It defines a general framework to fulfil the requirements of the recipient. Scientific needs, data availability and final user needs were analysed together to find a balance that optimised the quality of the collected data and the potential use of this data both within the project and by government. The environmental model resulted in a geodatabase that was the principal tool for execution of the project.

2 Materials and methods

2.1 Remote sensing

To perform the interpretation, a multiple-view remote sensing approach (LILLESAND and KIEFER, 2000) was selected combining multi-stage, multi-spectral and multi-temporal sensing. The process of convergence of evidence was used to successfully increase the accuracy and detail of the land-cover interpretations. For any application of remote sensing, not only the right combination of data acquisition and data interpretation techniques must be selected, but the right mixture of remote sensing with conventional techniques must also be identified. Remote sensing is a tool; it is not an end in itself.

In Mozambique data availability and quality are the main constraints to environmental and socio-economical studies. Even if the situation is quickly improving, base data such as topographic and geological maps, aerial photographs, etc. when available, are often out-dated or characterized by low quality informative content and positional accuracy. Limited budgets and logistic restrictions often limit *ad hoc* data collection, restricting projects to use to the few existing, available and reliable datasets.

On the other hand, land-cover interpretation is a complex task that involves spatial, temporal and spectral analysis. It is necessary to have data acquired for different seasons and different years in order to enhance the quality of the interpretation. A wide range of sensors with geographic, temporal, spectral and radiometric properties is available. Some data sources are free of charge, making their acquisition and use especially suited to developing countries.

The land-cover interpretation was primarily based on Landsat 5 TM images of the years 2004–2005 (39 scenes) (JANSEN et al., 2006). They were acquired at processing level 1G with Cubic Convolution re-sampling in format BSQ and georeferenced in the appropriate coordinate system (e.g., UTM36 South or UTM37 South, datum WGS84) and converted to geoTiff format. Since the number of Landsat scenes to be handled was considerable and the topographic maps available were from 1960's, with a very low positional accuracy, an alternative georeferencing method for the collection of Ground Control Points on topo-maps, providing both the required positional precision and not consuming too much time, was desirable. The solution adopted was an image-to-image georeferencing process using the NASA's Global Orthorectified Landsat Data Set as reference (TUCKER et al., 2004). These images from around the year 2000 have a documented positional accuracy of less than 50 m Root-Mean-Square Error (RMSE) which is adequate for the interpretation objectives of the Project. These images are freely available on the Internet from the Global Land-cover Facility web site (http://glcf.umiacs.umd.edu/).

These images were also used with the other ancillary dataset, to support the Landsat images interpretation. In Manica and Maputo Provinces, two or three images, depending on the availability, of the same scene from Landsat 7 ETM+ (circa 2000) were downloaded to comprise the seasonal variability of land-cover and one scene from Landsat 5 TM (circa 1990s) to analyse long-term temporal changes. At the national level, just one image per scene was downloaded because of the less detailed level of land-cover interpretation required and the subsequent impossibility to analyse a considerable number of satellite images in detail. 24 Landsat 5 TM and 64 Landsat 7 ETM+ were used.

A large dataset of ASTER VNIR (visible and near-infrared) scenes was downloaded from http://asterweb.jpl.nasa.gov/ (level IIb correction). The images are already georeferenced but the accuracy is low (e.g., around 100–50 m), thus these images were never used to assess the geometry of any land-cover polygon. The resolution of 15 m for three bands (corresponding to bands 2, 3 and 4 of Landsat) allows recognition of features and patterns with high spatial definition. ASTER images were used as a valuable support to define land-cover classes inside the polygons delineated using the Landsat 5 images in Manica and Maputo Provinces. They were used to observe land-cover in a season not covered by other images and to discover spatial patterns hidden in the Landsat images. At national level, ASTER images were used for validation of the land-cover interpretation. A total number of 71 scenes were used.

The reference scale of 1:250 000 (i.e., with the same metric and informative quality level of a traditional hardcopy map at that scale) for the provincial level corresponds with the use of Landsat images, widely used for land-cover data collection in scales ranging from 1:50 000 to 1:250 000. The reference scale of 1:1 000 000 to be used at the national level does not correspond with the use of Landsat images because they contain much more detail than was required. The possibility of using MODIS images for the national level was examined with the combination of the two bands at 250 m resolution (corresponding to band 3 and band 4 of Landsat) and a band at 500 m resolution (band 2 of Landsat). Since these results were not satisfactory because of the coarse resolution and the loss of contextual information of land-cover types, it was decided not to use MODIS data for land-cover interpretation. Therefore, Landsat images were re-sampled/degraded from a pixel size of 30 m to a pixel size of 120 m for land-cover type interpretation and 210 m for delineation of the polygons in order to lose detail in the image that does not serve the level of interpretation required, keeping all the desired contextual information. NDVI analysis based on MODIS and Landsat images were used to distinguish vegetated and non-vegetated areas in mixed classes and to quantitatively estimate their percentage.

The interpretation process was constantly monitored for internal consistency to ensure data quality control using the tools offered by the geodatabase: the spatial coherence of the polygon, i.e. were the boundaries in the appropriate place and had the same logical and functional thinking been applied in a consistent manner in the area of interpretation; the thematic coherence, i.e. was the label given to the polygon correctly describing its contents and were other areas with similar features (e.g., spectral signature) described in the same manner; topologic coherence, i.e. the spatial relationships among features were coherent with the defined topological rules.

2.2 Field survey: orientation and validation

It is possible to obtain land-use information in addition to land-cover when interpreting satellite imagery (CIHLAR and JANSEN, 2001). While land-cover is based primarily on spectral data content, which makes the successful use of digital image analysis possible, land-use information benefits from other aspects of the data including pattern, shape, context, size, shadows, etc. This contextual information may be efficiently extracted using visual analysis.

The land-cover interpretations based upon the Landsat images are integrated with field survey data (e.g., field orientation and field validation), data describing factors influencing land-cover (e.g., human activity, climate, terrain morphology, flora and fauna, soils) and with data on other land-cover aspects that are either coded (e.g., other land-cover data sets having different legends, dates and geographic detail) or uncoded (e.g., satellite images from ASTER, Landsat and MODIS covering different years and different seasons).

The application of remote sensing techniques cannot produce reliable results if there is not an investment made in establishing the relationships between what one can visually interpret on-screen and what one observes in the field. Therefore, two types of field surveys were executed for the two Provinces. Field orientation (111 observations) established a better understanding between the spectral reflectance values of the satellite images and the ground truth during interpretation, whereas the field validation (313 observations) confronted the interpretation with the ground truth in order to assess the thematic accuracy of the interpretation. These field surveys contributed to the high quality of the resulting land-cover data sets. For the national level, considerations of the costs/benefits of field surveys covering the whole of Mozambique should be balanced against one another. Therefore at the national level no further field surveys were executed but 338 observations on ASTER images having 15 m resolution served as validation data.

2.3 Ancillary data used

Land-cover is the product of the interaction between many factors, both anthropogenic and natural, thus it is very important to analyse ancillary data other than remotely sensed images. This additional data facilitates the understanding of land dynamics and brings the abstract qualities of land-cover interpretation closer to the real world.

Topographic maps are a fundamental instrument for the collection of land information. The cover of Manica and Maputo Provinces was available at 1:50 000 and 1:250 000 scales.

A Digital Elevation Model (DEM) is important to explore the relations between land-cover/use and morphology. The DEM from the Shuttle Radar Topography Mission (SRTM) was used. It has a resolution of 90 m and a standard quality (see http://srtm.usgs.gov/) that met the project objectives. Slope, hill-shade and aspect were derived from the original data set. It was also used for 3D rendering.

A contribution to facilitate satellite image interpretation was provided by the land-cover/use data set created by the IGN/CENACARTA/DINAGECA (1999) Joint Venture in the framework of the Rural Rehabilitation Project (1996–1999).

Other environmental databases used are: SOTERSAF, derived from SOTER and developed by FAO/ISRIC; UNESCO/AETFAT/UNSO Vegetation Map of Africa. This data was used for an approximate identification of Miombo and Mopane vegetation zones and for analysis of macro-ecological zones at national level.

A wide range of field survey data were used to support and confirm image interpretation results.

2.4 Land-cover legends

The AIFM land-cover legends were created, using the classes defined by the final users, through the FAO/UNEP Land-cover Classification System (LCCS). It allows a standardised, systematic definition of classes using a set of parameters tailored to the specific land-cover feature being described, harmonizing the final products in the broader context of natural resources management and regional planning. In Manica Province 31 and in Maputo Province 33 different land-cover classes were identified and at the national level 20 land-cover types were distinguished. There is a hierarchic link between the classes at the two levels. The LCCS is used by other international land-cover/use assessment programmes in Developing Countries (i.e., FAO AFRICOVER project).

The application of the legend to on-screen visual image interpretation required a good knowledge of the geographic area. The synergy between UIF and the executing company was therefore of the utmost importance.

2.5 Minimum mapping units applied

The applied minimum mapping units are 1 km^2 for the provincial level and 16 km^2 for the national level. A set of information is associated with the land-cover polygons. At the provincial level, the class of the dominant and secondary (if presents) land-cover is presented with their relative percentages. For the national level, a tertiary land-cover class (if present) and its percentage inside the polygon are added to the dominant and secondary ones.

2.6 The software used

The GIS software used was ArcGIS and SQL Server. ArcSDE handled the connection between the workstation GIS tools and the geodatabase located on the server. Erdas Imagine and GRASS were used for image processing.

2.7 The conceptual framework for the geo-database

It is important to define a conceptual framework for the land-cover/use assessment and change analysis, integrating different geographic levels and methodological approaches. As no environmental model of land-cover/use suitable for the AIFM project was found in the literature, an attempt to define a general environmental model was performed at the very beginning of the project.

Land-cover is a complex concept that involves both the definition of a set of (bio) physical cover characteristics to be considered and an instrument (e.g., human eye, digital sensor, etc.) to observe the Earth's surface, together with an observation point (e.g., plane, satellite, ground, etc.). How the observation of the (bio) physical cover on the Earth's surface will look like, depends on the instrument(s) used for observation, the time of observation, the location, the set and detail of (bio) physical characteristics observed and the geographic accuracy of the observation. Consequently, there are many possible representations of the same land-cover.

A general, approximate environmental model of land-cover can identify five main categories of interdependent factors that determine the (bio) physical cover on the Earth's surface: (1) morphology (e.g., elevation, slope, aspect, etc.); (2) human action (e.g., infrastructures, pollution, urbanisation, etc.); (3) climate (e.g., precipitation, temperature, etc.); (4) soil and geology (e.g., texture, rock outcrops, etc.); and (5) biotic factors (e.g., vegetation and fauna). All these factors are a function of space and time.

Land-cover data sets are real observations of the (bio) physical cover on the Earth's surface. They can be divided in two broad categories: (1) coded and (2) uncoded representations. Uncoded representations are, for example, satellite images (e.g., Landsat, Aster, MODIS) or vegetation indexes (e.g., NDVI). They provide a quantitative measure of physical properties (e.g., spectral reflectance, temperature, etc.) of land-cover. In coded representations some kind of (bio) physical cover classification system is used. An example of a coded representation is the Vegetation Map of Africa. Data collected in the field can be either coded or uncoded. This general overview is helpful as a conceptual framework where the data model and geodatabase structure have been developed (Fig. 1).

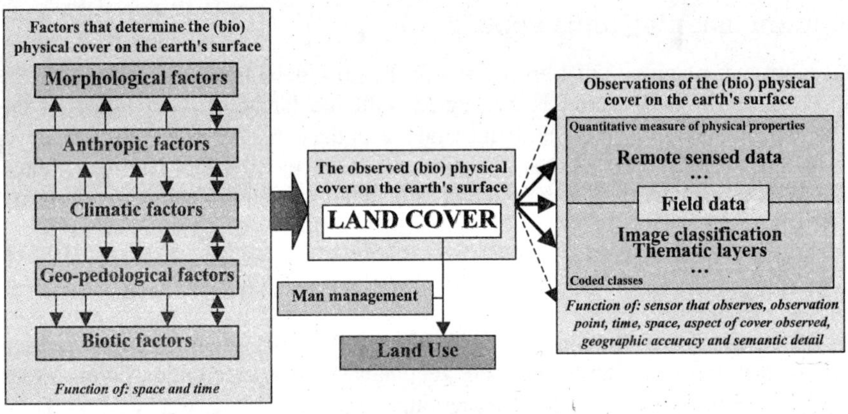

Fig. 1: General land-cover environmental model used as a framework by the project

The land-cover/use data set required by the AIFM Project is described in the Technical Proposal and Inception Report (TOR), in particular the classes to be incorporated in the land-cover interpretation at the two geographic levels (provincial and national), the thematic accuracy to be reached and the sensor to be used. There are many possible methodologies to develop the requested land-cover/use data set. One possible approach could be the definition of a functional relation between land-cover/use classes and the (bio) physical cover genetic factors. But due to the very complex nature of the classification requirements of the Project, it would be very hard to define such a model for the whole of Mozambique. Moreover, the datasets available in Mozambique are not always comprehensive, so that this kind of approach is impractical. Another possible methodology could be the construction of a predictive model based on statistical and geostatistical relations between classes of land-cover/use and a set of available variables related to land-cover/use. A number of points with known values for the variable to be predicted could be used to calibrate and validate this kind of model. Again, the semantic detail (i.e. classification) required makes this approach unsuitable and results are likely to be insufficient for an applied project whose main goal is not merely research. The methodology selected and applied for the collection of land-cover data is on-screen visual interpretation of Landsat images of the years 2004/2005, integrated with available digital ancillary data sets, field validation and accuracy assessments. To get this result, a GIS supporting and improving the satellite image based land-cover interpretation activities was developed. The geodatabase was the main tool of the GIS.

The use of GIS often takes place without intense dialogue with the beneficiary resulting in the implementation of a system that does not fulfil the needs of the user. To overcome these problems, an analysis of the system requirements was made at the very beginning to define jointly with the beneficiary the data to be included and the functions the information system should have. This requirement analysis, together with the land-cover/use conceptual framework discussed before and the project constraints (data availability, time, resources), led to the development of a data model implemented in a geodatabase, which is the heart of the information system. The data modelling effort was focused on the key purpose of accommodating land-cover and related data layers (e.g., land-cover change, field sample data, and forest type information).

The geodatabase is structured around five key thematic and functional groups (Fig. 2): (1) Land-cover, (2) Field Sample Data, (3) Ancillary Data, (4) Work Management and (5) Satellite Data. It relies on the tools offered by a classic relational database (CERI et al., 1999) and is organized using the structures offered by ArcGIS (ZEILER, 2003; ARCTUR and ZEILER, 2004): feature classes, domains, feature data sets, topological associations, table, relationships, raster catalogues and raster datasets. Metadata are associated to all data layers.

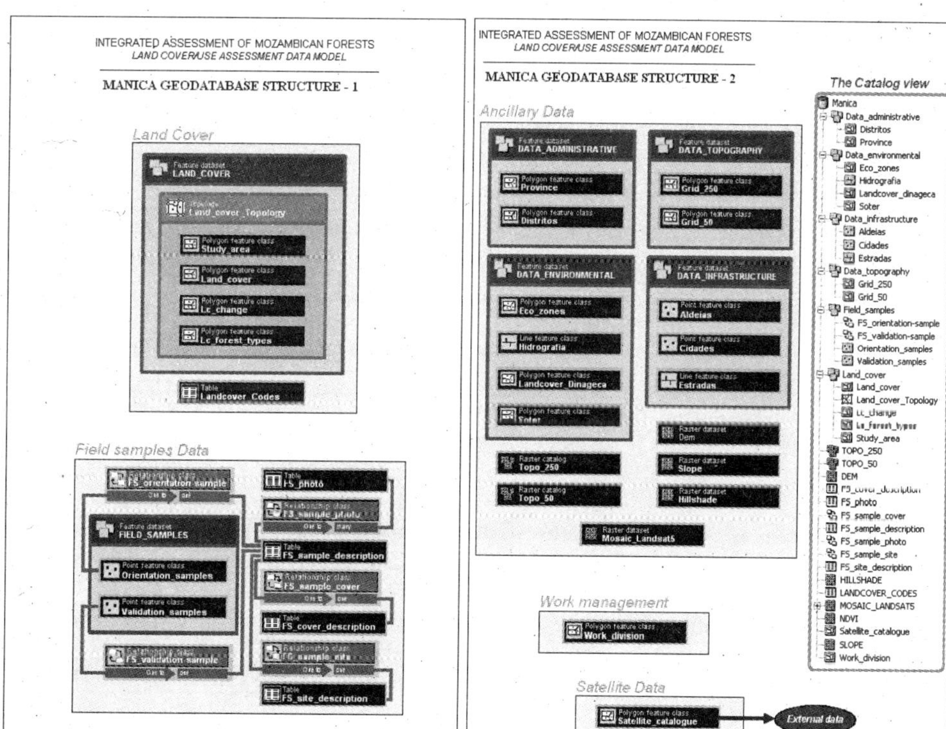

Fig. 2: Geodatabase structure

The geodatabase was designed to enhance integrity and consistency, both for geometry and semantic attributes and to permit editing of the same data set by multiple users at the same time. It defines topological rules to be respected (i.e. no overlapping polygons, no gaps among polygons, etc.) and attribute domains, thereby increasing the consistency of interpretation between users and overcoming edge-matching problems in areas where different satellite image interpreters are at work. The permission policy maintained data integrity even with multiple users working on the same geodatabase, with a GIS administrator supervising all data flows and overall data management.

3 Results and discussion

The outputs produced are land-cover data for Manica and Maputo Provinces for the year 2004 (at a reference scale of 1:250 000) and national land-cover data (at a reference scale of 1:1 000 00) for the year 2004–2005, outlining the extent, distribution and relationships among different land-cover types in digital format. The thematic accuracies of the land-cover data for Manica Province, Maputo Province and Mozambique are 87.0%, 86.0% and 86.0% respectively. The positional and thematic accuracy assessment results allow the use of these data sets at the required planning and management level. Table 1 summarizes land-cover statistics at the national level.

Table 1: Statistics for the national level land-cover interpretation for Mozambique

LCCS land-cover category	Land-cover class		Area	
	User code	User name	(ha)	%
A11. Cultivated and managed terrestrial areas	TC	Tree crops	3 285 954	2,24
	1FC	Field crops	13 477 171	9,18
	1CXF	Shifting cultivation with forest	5 212 027	3,55
A12. (Semi-) natural terrestrial vegetation	2FE	(Semi-) evergreen forests	9 728 145	6,63
	2FD	(Semi-) deciduous forests	32 161 583	21,90
	2WE	(Semi-) evergreen woodlands	1 734 117	1,18
	2WD	(Semi-) deciduous woodlands	28 929 290	19,70
	2TK	Thickets	2 533 778	1,73
	2SL	Shrublands	14 912 522	10,16
	2GL	Grasslands	14 684 820	10,00
	2FXC	Forest with shifting cultivation	7 015 031	4,78
A23. Cultivated aquatic areas	3AC	Cultivated aquatic areas	14 772	0,01
A24. (Semi-) natural aquatic vegetation	4FF	Aquatic forests	689 237	0,47
	4WF	Aquatic woodlands	1 472 900	1,00
	4SF	Aquatic shrublands	1 088 803	0,74
	4HF	Aquatic herbaceous vegetation	4 342 594	2,96
B15. Artificial surfaces and associated areas	5BU	Built-up areas	1 862 299	1,27
B16. Bare areas	6BA	Bare areas	1 877 652	1,28
B27. Artificial water bodies	7WB	Artificial water bodies	46 954	0,03
B28. Natural water bodies	8WB	Natural water bodies	1 765 028	1,20

A land-cover change analysis between 1990 and 2004 was carried out in Manica Province, where a total of 12.6% of the surface underwent land-cover changes. A complete description of the results can be found in JANSEN et al. (2006).

A land-cover/use assessment geodatabase was created and populated with a large set of environmental data, so that UIF/DNFFB can use it as a framework to build on and include new land-cover and other data sets in future.

The option to use on-screen visual interpretation at the national level based on Landsat images should be revisited. Moreover, the national level is more suited for a different kind of analysis, based on a general overview of the country and its changes, where broader classes are considered and automatic classification methods are used on images from sensor with a lower spatial but high temporal resolution (i.e. MODIS).

From a methodological point of view, even if a general environmental model can be defined to establish a framework for land-cover assessment, the development of an interpretation methodology and a data model is strongly dependant on the specific conditions of each project (objectives, data availability, time and resources, land-cover classes required, extension of study area etc.).

References

ARCTUR, D. and ZEILER, M., 2004, Designing geodatabases: case studies in GIS data modelling. ESRI Press, Redlands, California, USA.

CERI, S., ATZENI, P., PARABOSCHO, S. and TORLONE, R., 1999, Basi di dati. McGraw-Hill Publishers, Milan, Italy.

CIHLAR, J. and JANSEN, L.J.M., 2001, From land-cover to land-use: a methodology for efficient land-use mapping over large areas. The Professional Geographer, 53(2), pp. 275–289.

DI GREGORIO, A. and JANSEN, L.J.M., 2000, Land-cover Classification System (LCCS): classification concepts and-user manual. FAO, Rome.

IGN/CENACARTA/DINAGECA JOINT VENTURE, 1999, Land-use/land-cover legend.

JANSEN, L.J.M., BAGNOLI, M., CHIMENTI, S., DELL'AGNELLO, A., FOCACCI, M., MONACO, S., SORINI, P., STOPPIONI, S. and URBANO, F., 2006, Satellite image interpretation of land-cover types in Manica and Maputo Provinces at nominal scale of 1:250 000 and at national level at nominal scale of 1:1 000 000. Technical Report. Florence, Italy.

LILLESAND, T.M. and KIEFER, R.W., 2000, Remote sensing and image interpretation. Fourth edition. John Wiley & Sons, New York.

TUCKER, C.J., GRANT, D.M. and DYKSTRA, J.D., 2004, NASA's global orthorectified Landsat data set. Photogrammetic Engineering & Remote Sensing, 70(3), pp. 313–322.

ZEILER, M., 2003, Modelling our world. ESRI Press, Redlands, California, USA.

Geo-Information Management for Agricultural High Value Product Supply Chains

Norbert NIEDERHAUSER, Walter RITTER, Thomas OBERTHÜR
and Clemens BERTSCHLER

Summary

Developing high-value-crop supply chains such as coffee has the potential to improve the livelihoods of poor rural farmers in less developed countries. Systematic information management including spatial data can greatly facilitate supply chain development for example by linking a specialty market in one country with potential producers in another.

In this paper we describe the development of a prototype information management system for specialty coffee supply chain participants. All information in this system is linked to geo-referenced data (GPS data). This enables spatial data analysis in a local context to facilitate rural development. By linking certain product characteristics (e.g. coffee quality) in real time to existing spatial models (altitude model, climate, political division) certain analyses will immediately reveal specialty coffee production-niches. This information linkage also allows for analysis of product consistency in a spatial context e.g. regarding quality.

A key challenge is to facilitate interaction between rural coffee farmers and centralized information systems by bridging the existing technological gap. In this paper we also present results of a field-study about the design of optimized human-computer interfaces conducted in February 2006 with 45 coffee farmers in rural areas of Southern Colombia.

1 Introduction

Millions of farmers in developing countries are facing a downward spiral of poverty, driven by factors beyond their control such as price collapse, climate change, natural disasters or unfair trade environments. In many cases, farmers could break out of this downward trend by adaptation, for example, by diversifying into higher value crops or other alternatives. (UNITED NATIONS, 2003; FRITSCH, 2002; DICUM and LUTTINGER, 1999; RICE and MCLEAN 1999). Small-scale farmers, which make up 70 percent of the world's coffee growers, lack market information that is needed to make informed decisions about production. (DANIELS and PETCHERS, 2005; RAWORTH et al., 2002). Furthermore they need to know about the right time, quantity and location to grow quality coffee but lack any environmental information and support as well as they hardly get feedback about the quality of their coffee. In contrast with the mainstream price slump in the commodity market, niche or speciality coffees have been a runaway success that let farmers benefit by branding the quality coffee from their country or its origin. (RAWORTH et al., 2002) The key for entering the speciality market is information. Producers need site-specific information that is specific to their farm's situation as well as market information about consumer demands and trends (DANIELS and PETCHERS, 2005; COOK et al., 2004). For successful agricultural supply chains technologies for product identification, process and environmental characterization, information capture, analysis, storage and transmission as well as overall system integration are necessary. Innovations in geospatial science, like geographic information systems (GIS), remote sensing (RS) and global positioning systems (GPS), offer great opportunities for site-specific agriculture and the derivation of data related to the agriculture products, on the farm and through the subsequent handling operations. These technologies enable furthermore the option to map the geospatial variability of selected attributes such as yield, product quality or movement and disease epidemiology. Radio frequency technology (RFID) and mobile tracking devices are able to collect and transmit data from tags to distant locations for storage and analysis. The simplest label stores information like name of product, lot number and price but also data about product origin, handling and storage conditions and other quality attributes like for example a taste profile could be stored (OPARA, 2003; STAFFORD, 2000). As coffee quality is dependent on many factors (environmental and genetic factors, agronomic production, post-harvest management), an geo-information management system is needed that collects all information upstream of the supply chain and supplies it downstream to the traders and customers (GOLDSMITH and BENDER, 2004). Furthermore the geo-information management system links the collected information to geo-referenced data (GPS data). This enables spatial data analysis in a local context to facilitate rural development.

Therefore this paper focuses on the prototype of a geo-information management system for agricultural high value products, CinfO, which is explained by the case study of speciality coffee. As farmers have been the only supply chain participants so far which have been excluded from the supply chain system because of lack of access and knowledge of the according technology, a field study has been conducted on how to link farmers in by using the mobile computing technology. Results of this field-study about the design of optimized human-computer interfaces conducted in February 2006 with 45 coffee farmers in rural areas of Southern Colombia are presented in this paper.

2 CinfO – Prototype geo-information management system for agricultural high value product supply chains

2.1 System-overview

The prototype CinfO system consists of two parts: a central database and an Internet based user interface. Surveys were carried out to determine users' data requirements. A multi-user framework tailors information to users' needs in respect to data privacy by granting users specific access rights, and providing support for language and visual needs. The prototype CinfO incorporates information about coffee production, farm management and coffee quality. All of this data is geo-referenced which allows for a later spatial analyses and visualization on maps. A unique coding system guides the data along the supply chain and guarantees traceability. The whole system is based on Internet technology, which enables quick and easy input, update and feedback of information of all participants in the supply chain in real-time. This is crucial in a highly dynamic market with constantly changing demands. A simple web-mapping interface allows for querying the database for certain information and presents the results on the web-map as well as in textual and graph form.

By this visualization module all farm- and production data become more accessible and comparable (NIEDERHAUSER, 2004).

2.2 The CinfO product tracking system

One main goal of the CinfO system is to identify both production quality and product characteristics (e.g. flavor) and to link this information to the final product through all levels of product tracking. The product-related information is linked to GIS data (geo-reference coordinates given longitude, latitude and altitude), which enables interactive mapping of product and production data on digital maps via the online interface (NIEDERHAUSER, 2004).

In the case of coffee the product track runs from the field where coffee is grown to consumers all over the world while the physical state of coffee changes from the green, fruit pulp-covered coffee bean to a wide variety of roasted, blended, grinded and packed coffee of different flavours for different use. The main product tracking nodes through which coffee runs are: production on the field, post harvest processing on the farm (fruit pulp peeling, fermentation, and washing, drying), further processing in coffee cooperatives (sorting, quality check, peeling and others), changing of ownership and transportation through exporters to importers to roasters (roasting, blending, grinding), and finally the marketing process which results in the end-point sale to a customer (general public or food industry) (NIEDERHAUSER, 2004).

Data collected for the CinfO system include the management unit (e.g. fields) that include environmental data about soils, climate and geography (e.g. size, slope, GIS data), coffee crop information, including management data (planting, shade trees, crop combinations), farm management data (farming technology for harvesting and product processing), human resources, farm management costs and coffee quality data (flavour, coffee diseases and humidity) (NIEDERHAUSER, 2004).

Static data (e.g. aerial photographs of fields and farm, GIS coordinates, soil and climatic data) will be captured and entered into the system by CIAT staff due to the need for using specific capturing tools and methods. Frequently captured data (production, costs, laboratory quality test data....) should be decentralized by giving the responsibility to farmer associations, trusted and competent local organizations and all other participants. To ensure a certain level of data quality, common standards on measuring and capturing data must be applied.

To link final products to origin information a set of product tracking codes are applied. These codes are directly attached to the product and enable participants to obtain information about the product from the CinfO database that serves as central data storage module (NIEDERHAUSER, 2004).

2.2.1 Code systems for product tracking

CinfO distinguishes three different code systems (Fig. 1), the management unit code, the harvest code and the product code. The management unit code is used for linking all production and processing data as well as geographic information about the production zones to a specific environment, defined as management unit. The harvest code combines the management unit code with the harvest date which enables the tracking of quality data and other harvest dependent parameters. The product code is mainly for end customer use to identify the origin of the packed and ready-to-sell product. Due to the fact that many coffee roasters use a mixture of coffee from different regions and producers, this code is needed to link several production codes with the help of a new code that appears on the coffee pack. (NIEDERHAUSER, 2004)

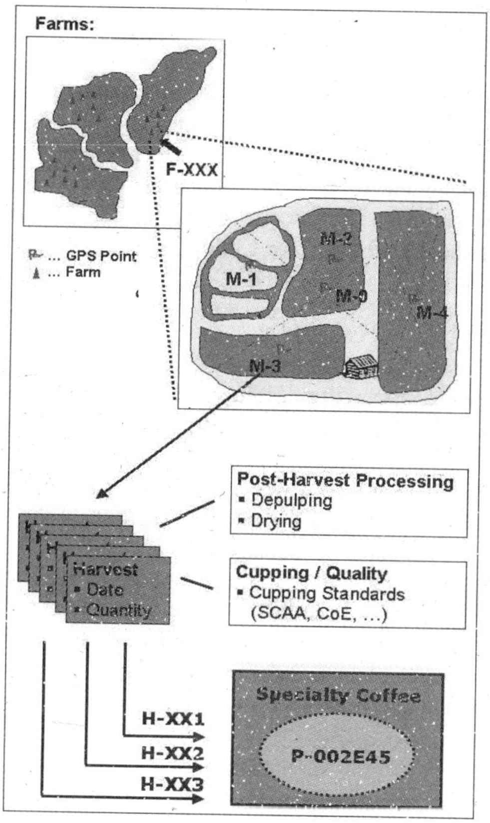

Management-Unit-Code:
[Example: **M-0078F6**]
- Generated on management-unit registration in database.
- Enables to link site-information in database.

Code users: producers, coffee coöperations

Harvest-Code:
[Example: **H-0010B0**]
- Generated on harvest registration in database.
- Links harvest date and management-unit-code.

Code users: coffee cooperations, coffee quality laboratories, exporters, importers, roasters

Product-Code:
[Example: **P-00006F**]
- Generated from roaster on registering end-product in database.
- Links the different harvest-codes of one end-product.

Fig. 1: Basic code system enabling product tracking along the supply chain

The management unit code has to be a simple code, easy to read and process. Furthermore it must be unique system-wide and equipped with a simple error detection mechanism. To provide a sufficient and easy to implement error detection mechanism the checksum method was selected. On registering a new field in the CinfO system, a new system-wide unique number is generated. Afterwards a checksum bit is added to this number and the whole expression is converted to a hexadecimal number. Finally the expression is shorter and the alphanumeric presentation is easier to read for humans. TANENBAUM and GOODMAN (1999) provide detailed information about the checksum algorithm.

For the harvest and product code the same principals as for the management unit code are applied. The harvest code is created by adding a new artificial code using a table index of the database. The prefix enables an easy determination of a given code. For creating the product code two different possibilities exist. The roaster can either select the harvest codes for all raw coffees used for a mixture and receives a product code from the CinfO system, or the roaster uses eventually the existing product bar code and links this code with the production codes existing in the CinfO database (NIEDERHAUSER, 2004).

WAGNER and GLASSHEIMER (2003) explain a similar but more generic approach for a food tracking system.

2.2.2 Code presentation

As mentioned above, the code is presented in hexadecimal numbers. Since electronic scan devices and RFID technology are not available for the CinfO pilot project phase, this is a considerable advantage. If barcode printers and scanners or the RFID technology are available in the future, it would be easy to implement them within this numeric system. A popular standard for presenting text and numbers as barcode is the Code39 (also commonly called LOGMARS, Code 3 of 9 or the 3 of 9 Code) or Code 128 (also referred to as ISBT 12, UCC-128, EAN-128 and USS Code 128) (BARCODE, 2004).

2.3 Data presentation

The CinfO platform provides general information about the coffee tracking project and specific information about coffee production in the pilot regions including different queries that enable simple statistical analyses and an information presentation interface adapted to specific user-groups. The specific information section provides each user-group with detailed coffee production and quality information on a specific farm including management information, soil and climate information, coffee flavour and coffee production information as well as historic data about quality. CinfO offers a highly flexible structure for managing the dynamic and sensitive content tailored to the user's demand giving them access rights only for specific information that the user is allowed to see. The data access model is realized through a three level security schema dividing the CinfO platform into a public, private and administration area (NIEDERHAUSER, 2004).

2.3.1 CinfO online map interface

A special interface enables the consumer to receive information about the coffee's origin by entering the product code shown on the coffee pack including all farm related data where the coffee was grown. The Minnesota MapServer is used to access and visualize location-based knowledge. The user retrieves the required information by using simple selection or search tools with the options to search by farm or farmer organisation, by region, country or 'departamento' or by area/farms where certain product characteristics are met. (Fig. 2, section 1) Different interaction options with the map are provided through zooming, panning or creating self-defined maps with the preferred spatial information concerning cities or rivers. The resulting map simplifies the spatial information and furthermore gives the possibility to extract additional data from the map by clicking on a specific farm. (Fig. 2, section 2) This data includes general farm data, GIS data and quality data (Fig. 2, section 3) (NIEDERHAUSER, 2004).

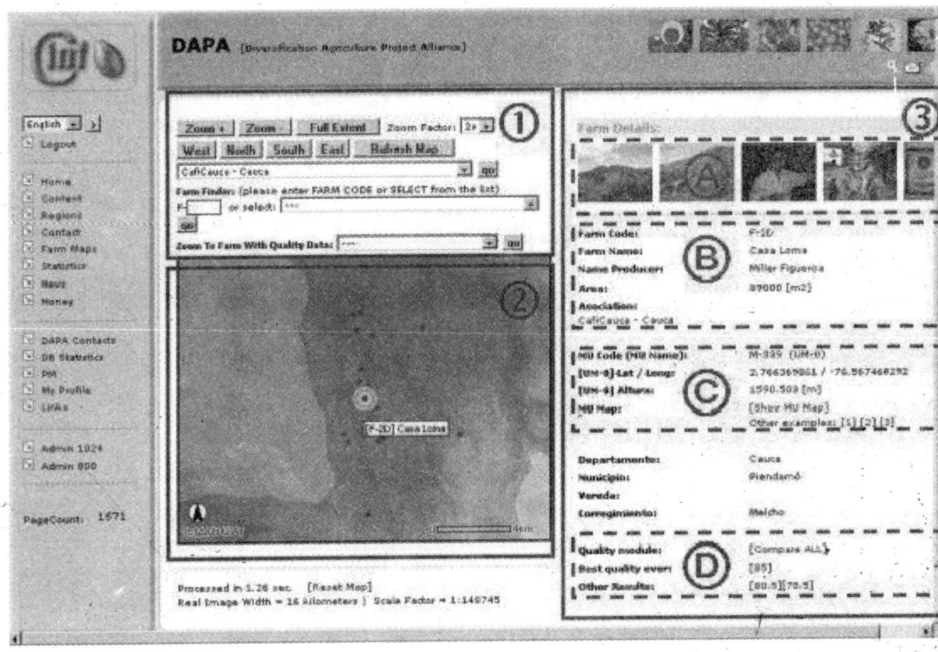

Fig. 2: CinfO Online Map Interface

2.3.2 Quality module

Basic visual data analysis, as shown in figure 3, is also possible through this module. Data can be queried (selection of query items and input of threshold values) and visualized via this map module. The analyst can see certain trends in respect to e.g. coffee quality linked to spatial or environmental factors (NIEDERHAUSER, 2004).

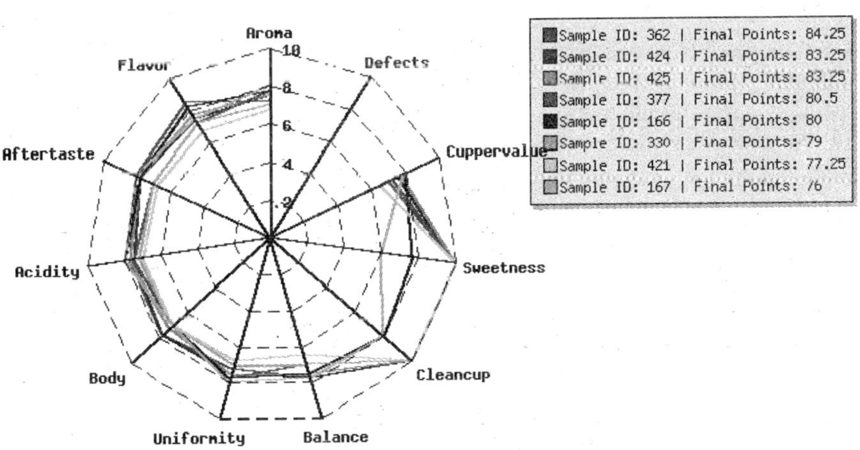

Fig. 3: Coffee Quality Module

2.3.3 Feedback-module

An intelligent feedback system provides the farmer for example with a quick overview of his productivity status. Automated processes for complex data analysis (involving forecast models) are producing easily understandable and powerful reports in different presentation. An example is demonstrated in figure 4, a farm map that indicates the productivity status of different management units (NIEDERHAUSER, 2004).

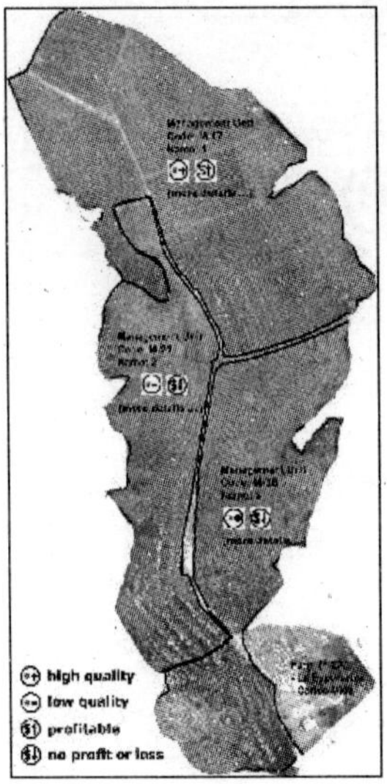

Fig. 4: Aerial photograph of farm management units and their rating in regard to production and quality

2.3.4 Spatial analyses and data export

By linking certain product characteristics (e.g. coffee quality) in real time to existing spatial models (altitude model, climate, political division) certain analyses will immediately discover specialty coffee production-niches. Any data stored in the central CinfO database is available as download-file in different formats: csv-file (comma separated value, readable with any spreadsheet program), shape-file, Google-Earth® format (Fig. 5) and a special file format for Expector – a software program for statistical analyses.

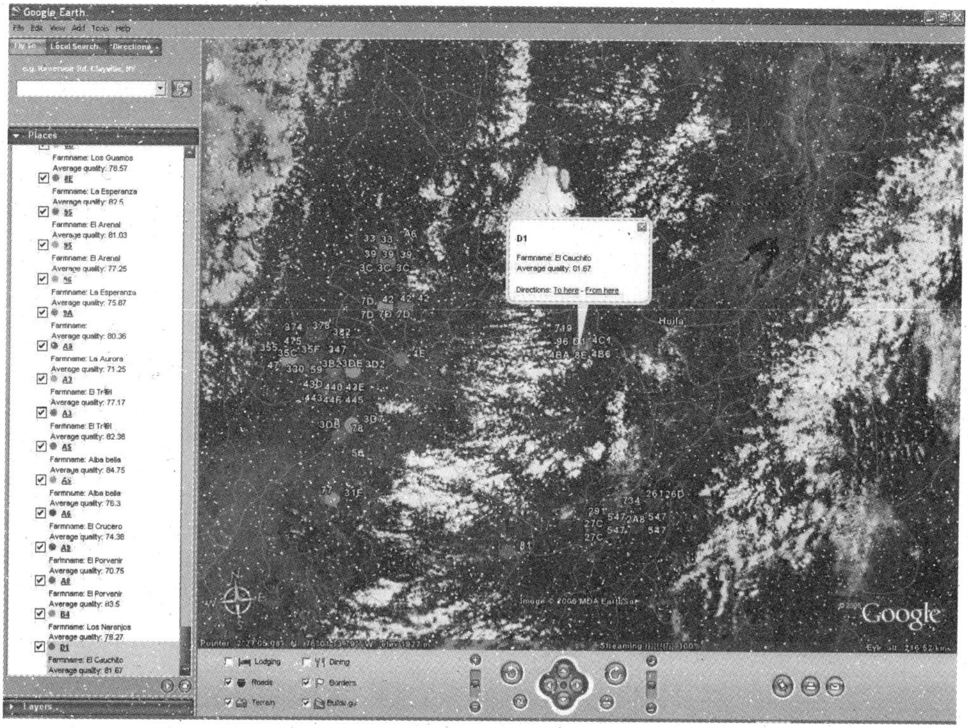

Fig. 5: Exported CinfO data can be used in Google-Earth®

3 Linking farmers in

The markets pursue for complete control over the supply chain, from the producer right to the consumer, also has its effects on rural development areas. Getting timely information about farm performances in these areas is essential for multiple reasons. Firstly, information about the quantity and quality of available products is important for accurate market predictions and therefore adequate and fair trading agreements. Secondly, farmers can benefit from valuable farm specific and time critical agricultural tips from research institutes like the Centro Internacional de Agricultura Tropical (CIAT). Finally, timely performance data from the field is also critical for research institutes trying to help farmers optimize their land use, and thus ensure better economic and ecological development. Currently, third parties collect such data manually, which is time consuming, expensive, and even risky at times. Given the advancements in the development of information appliances, both, technically as well as cost-wise, an online information system would be highly beneficial. However, introducing an easily accessible information system, besides technical issues like connectivity and device-robustness, involves crossing a huge barrier: people in rural areas seldom have any computer knowledge at all. Therefore intuitive operation of such a system is a critical factor for its acceptance und use, especially because comprehensive trainings are not an option due to high costs, time- and money-wise.

Since mobile phone providers are extending their networks into rural zones and (mobile) computing technology advances, new opportunities arise due to both, new kinds of devices as well as lowered prices. Newly designed and optimized human-computer interfaces will enable farmers to make use of portable computing devices such as PDAs. For implementing this lightweight technology in a user-centered way, a field study was conducted to gain relevant insights about how computer novice users like farmers in rural areas are getting along with standard input controls on PDAs. These results form the basis for the design and optimization of the user interface. Once farmers are able to manage their data independently, the last missing link to complete an autonomous functioning supply chain is established. The whole system will gain momentum and power.

Within the field-study we tried to get an idea of how open farmers in rural development areas are towards the use of handheld computers, evaluating how farmers get along with standard input controls like text fields, pull-down menus, or checkboxes, learning what makes it hard for the novice user to get along, and finally drawing some conclusions on how to create intuitive questionnaire-interfaces for small handheld devices. All these findings should assist the development and introduction of a mobile information appliance for information management in rural development areas. As collecting user data from the field is widely considered essential for successfully developing mobile applications (HOLTZBLATT, 2005; BLOM et al., 2005), we decided to conduct a field study in the environment where the mobile application is going to be used finally – in the coffee region of Huila, Colombia.

To find out what input controls can be used in an intuitive way, we decided to present four different electronic forms to test participants, with each of the forms featuring a different set of input controls. The participants should then enter their answers for the same four questions into these different forms (see figure 6 for screens of the four different form variations).

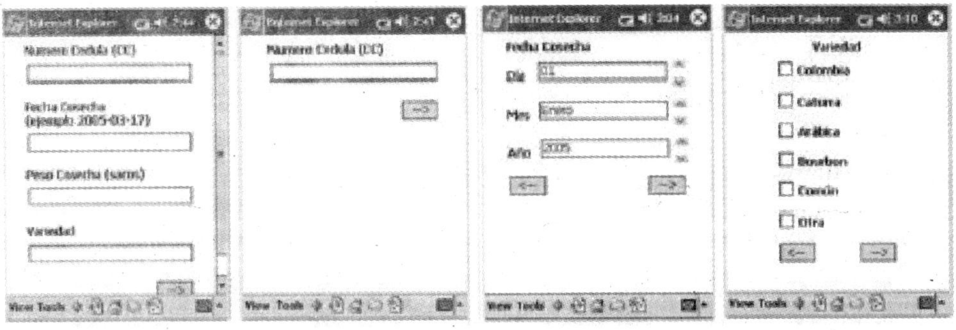

Fig. 6: Screens of the different form variations

Farmers first had to fill out a paper form with four questions. Then they were asked the same questions using four different digital form variations in four different sequences. After each completed form participant were asked questions and at the end they were asked for their final impressions. All in all 41 randomly selected persons participated in the tests (36 males, 5 females).

The study showed that farmers are very open towards handheld computers. They feel comfortable using digital forms and put more trust into this kind of technology. The results of the field-study are:

- Reduce varieties of controls to a minimum
- Avoid contextual menus showing up after long clicks
- A dedicated device without any distracting features would be favourable (at least full screen-mode)
- Stay consistent – within the forms themselves, as well as among different forms
- Optimized virtual keyboard development

Based on this study we suggest to develop a dedicated information appliance for gathering and viewing relevant information. Having a dedicated device makes the interface cleaner and thus easier to understand (compare NORMAN, 1999). The interface should be free of any distracting elements present for unused functionality. Our study also proved that consistency plays an important role for efficiency: sticking to monotony in available interface elements is critical. This importance has also been pointed out in various other places (RASKIN, 2000; SHNEIDERMAN, 1998; NIELSEN, 1993). Consistency within the application itself, as well as among different questionnaires should therefore have highest priority. We also propose that further tests should be performed with a virtual keyboard featuring a natural and therefore predictable letter order. This might help to improve text input speed for novice users.

4 Future research and challenge

According to the experience of supply chain members, there exist different views about what is coffee quality on different nodes along the supply chain. To give an example, for someone a coffee is of good quality when it doesn't smell like chemicals while another one associates good quality with elegant management. Future research is done to make the information flow along the supply chain with respect of different technologies available at certain nodes in the supply chain and prepare the information according to the knowledge and needs of different SC-members.

Future work attempts to diminish information asymmetries along the supply chain, to enable rapid information flow and product manipulation, and bridging the gap between humans and technology.

The challenges lie in overcoming the connectivity problem; create special user-computer interfaces for farmers, presentation of information and the easy usage of this information (Information Engineering).

5 Conclusion

This paper outlines the importance of geo-information management for entering the speciality channel of agricultural products and presents as a result the prototype geo information management system CinfO. Moving away from the commodity channel and entering the exclusivity or speciality channel could improve the live of millions of smallholder farmers (BARGHOUTI et al., 2004; INTERNATIONAL COFFEE ORGANIZATION, 2004), but sets high standards on quality and information processing. Producers need site-specific information that is specific to their farm situation to produce for a certain niche-market. This could only be achieved by using geospatial science, like geographic information systems (GIS), remote sensing (RS) and global positioning systems (GPS). Furthermore, these technologies enable the option to map the geospatial variability of selected attributes such as yield, product quality or movement and disease epidemiology. By linking the collected information along the supply chain to GIS data, spatial data analysis becomes available, that again enables certain analyses to discover speciality coffee production-niches. To complete the autonomous functioning supply chain, a field study was conducted how farmers could be linked to the system and the results are presented in this paper. Newly designed and optimized human-computer interfaces will enable farmers to make use of portable computing devices such as PDAs to replace the manual data collection process that is time consuming, expensive and at times risky.

References

BARCODE (Barcode Standards Organizations), 2004, http://www.makebarcode.com/stdorg/orgs.html.

BARGHOUTI, S., KANE, S., SORBY, K. and ALI, M., 2004, Agricultural Diversification for the Poor. Guidelines for Practitioners. Agricultural and Rural Development Discussion Paper 1. The International Banc for Reconstruction and Development Agriculture and Rural Development Department, pp. 1–48.

BLOM, J., CHIPCHASE, J. and LEHIKOINEN, J., 2005, Contextual and cultural challenges for user mobility research. Communications of the ACM, 48(7), pp. 37–41.

COOK, S.E., O'BRIEN, R., OBERTHUR, T. and CORNER, R.J., 2004, Is precision agriculture irrelevant to developing countries? CIAT-International Center for Tropical Agriculture, pp. 1–6

DANIELS, S. and PETCHERS, S., 2005, The coffee crisis continues. Situation assessment and policy recommendations for reducing poverty in the coffee sector. Oxfam America, pp.1–42.

DICUM, G. and LUTTINGER, N., 1999, The coffee book. New York Press.

FRITSCH, P, 2002, Robust coffee profits are not trickling down to Nicaraguan coffee growers. The Wall Street Journal Europe.

GOLDSMITH, P. and BENDER, K., 2004, Ten conversations about identity preservation. Journal on Chain and Network Science, 4, pp. 111–122.

INTERNATIONAL COFFEE ORGANIZATION, 2004, Diversification in coffee exporting countries. Review of past and present initiatives. http://www.ico.org/.

HOLTZBLATT, K., 2005, Designing for the mobile device: Experiences, challenges, and methods. Communications ofthe ACM, 48(7), pp. 33–35.

NIEDERHAUSER, N., 2004, Internet based product tracking and information system for coffee and other high value crops. Master thesis. iTec – information and communication engineering. University of Applied Sciences Vorarlberg, Austria.

NIELSEN, J., 1993, Usability Engineering. Academic Press, San Diego, CA.

NORMAN, D.A., 1999, The Invisible Computer. The MIT Press. Cambridge, MA.

OPARA, L.U., 2003, Traceability in agriculture and food supply chain: a review of basic concepts, technological implications, and future prospects. Food, Agriculture & Environment, 1(1), pp. 101–106.

RASKIN, J., 2000, The Humane Interface: New Directions for Designing Interactive Systems. Addison-Wesley, Reading, MA.

RAWORTH, K., WILSON, D. and HAWARD, B., 2002, Mugged poverty in your coffee cup. Oxfam International 2002, pp. 1–54

RICE, D. and MCLEAN, J., 1999, Sustainable coffee at the crossroads. White paper prepared for "the consumer's choice council".

SHNEIDERMAN, B., 1998, Designing the User Interface: Strategies for Effective Human-Computer Interaction. Addison-Wesley, Reading, MA, third edition.

STAFFORD, J.V., 2000, Implementing precision agriculture in the 21st century. Journal of Agricultural Engineering Research, 76, pp. 267–275

TANENBAUM, A.S. and GOODMAN, J., 1999, Computerarchitektur. Prentice Hall. 4th Edition. München, pp. 87–91.

UNITED NATIONS, 2003, Economic development in Africa. Trade performance and commodity dependence. United Nations conference on trade and development (UNCTAD). Geneva.

WAGNER, G.L. and GLASSHEIM, E., 2003, Traceability of Agricultural Products. Northern Great Plains Inc, pp. 1–18.

GIS Based Water Resource Management of the Dead Sea Region – Integrating GIS, System Dynamics and Agent Based Modelling

Ernst GEBETSROITHER and Wolfgang LOIBL

Summary

The paper presents a prototype of a simulation platform developed in JAVA which combines the two approaches System Dynamics and Agent Based modelling (ABM) and uses GIS data as input to calculate spatial development maps of the water and land management within the Dead Sea Basin. The ABM is based on land use change probability layers whose conductance is described in this volume (LOIBL and GEBETSROITHER, 2007). With this integration of the two approaches a top down system analysis could be linked with a bottom up analysis to evaluate each other. Thus top down policy decisions can be translated to spatially explicit results and reactions by spatially acting agents can be visualised and used to evaluate the policy decision.

1 Introduction

This paper presents preliminary results of the EU-project "A Future of the Dead Sea Basin" (*INCO-MED contract no ICA3-CT-2002-10019),* which has the overall objective to establish the scientific basis for a "more sustainable than today" water management and water-related land management in the Dead Sea basin, and from this, to develop practical recommendations that can be used for strategic decision making. GIS data together with a detailed system analysis was used to develop a regional spatial simulation model. This model provides development GIS-maps for the region for several different scenarios to decision makers. The developed approach combines a top down view, based on the *System Dynamics* Method and a bottom up simulation, based on *Multi Agent Based Modelling.* Therefore top down measures performed by decision makers can be combined with the micro behaviour of agents like inhabitants, farmers, or entrepreneurs establishing businesses or hotels. Feedback from the agents can be used to evaluate and adapt the top down measures to establish a more sustainable water management decision.

This paper concentrates on the integration of the two approaches and GIS visualization into one decision support tool.

1.1 System dynamics approach

The approach handles "reality" as a system of stocks and flows of resources (material, money etc.) considering the interdependencies and the reaction of the system's elements. An event-oriented linear view of the world from a state with goals, problems and decisions to results is replaced by thinking processes and feedback loops reflecting these interdependencies. Goals lead to decisions which change the environment and this in turn leads to new decisions. A simple example is the price of a product which often mainly influences the customer's decision if they buy it and how much is bought influences the price of the product in a feedback loop. But as we know, such feedback loops often have delays and seem non-existent not only because of the inherent delay. Furthermore it is often neglected, that even exponential growth in a short period of time seems to be linear. Without analysing the feedback structure, sometimes policy resistance arises (STERMAN, 2000). The main steps of the approach can be described as followed:

- Formulating a verbal description of the system
- Description of the problem
- Defining the system boundaries
- Identification of main driving forces
- Identification of main system elements (extracted from verbal description)
- Developing of Causal Loop Diagram
- Investigation of relations between system elements
- Qualitative analysis of system behaviour
- Development of Stock Flow Diagram
- Quantification of relations between system elements of SFD's
- Simulation of Computer model
- Quantitative analysis of system behaviour
- Verbal description of improved system understanding

To build the model, the system dynamics software[1] Vensim DSS has been applied. The model contains a complex model structure which on the one hand simulates the water supply development and on the other hand the water 'demand' development. The following chapter will briefly describe some parts of the model.

1.2 Structure of the system dynamic model (input for the GIS modelling)

As figure 1 shows, Vensim enables to create and visualize a system structure with different kinds of elements (boxes and flows). The model building environment allows to create the system structure graphically through boxes representing stocks and links (with certain equations) representing relations. In this way model structure and elements are transparent, allow easy adaptation and are visible for the model users instead of being hidden by computer code. Vensim provides further additional features to analyse the inter-dependencies and the causes-uses relations to explore the underlying system structure, or features as sensitivity analysis, reality checks, tables and graphs to observe the process and interfere by changing stock numbers and flow parameters. A freely available model reader allows to view, read and thus integrate the model into different applications.

Results from the system dynamics model e.g. the population development within different scenarios, were used as an input for the Agent Based Model to analyse where regional development will take place and whether the top down assumptions make sense. (e.g. whether there is enough suitable land available to fulfil all the different purposes (for dwellings, agriculture, tourism, industry and nature).

---

[1] http://www.vensim.com/.

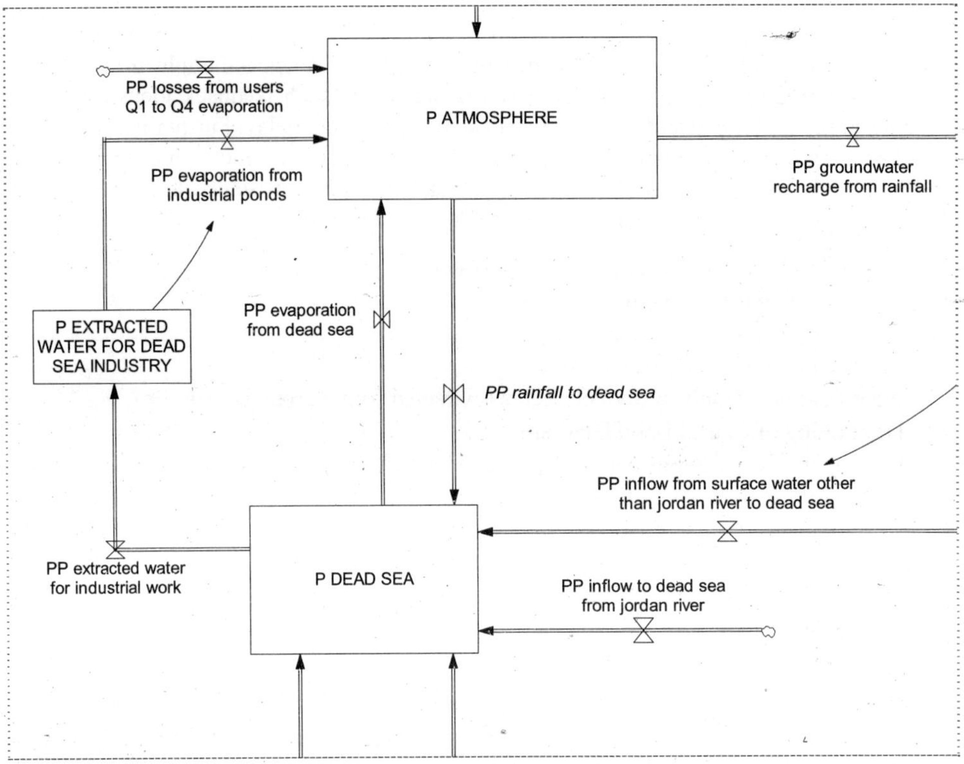

Fig. 1: Boxes and Flow structure of a Vensim model

2 Agent Based Modelling

Agent Based Modelling (ABM) has proven to be a remarkably versatile approach for understanding the behaviour of socio-ecological systems (GILBERT and CONTE, 1995; BERGER, 2001). It enables to address fundamental issues of the agency-structure debate, which highlights that social systems consist of both actors and structures/environment, each of which has an impact on the other (GIDDENS, 1979). In particular, an ABM can simulate interactions at micro level within water management systems (WMS) and its institutional environment. ABM gives special attention to co-evolutionary dynamics of the organisational levels (e.g. of water consumers and providers), which interact through rules managing the behaviour of agents. Recent successful applications of ABM in participatory water management (TILLMANN et al., 2001; BARRETEAU et al., 2003; BECU et al., 2003; DRAY et al., 2006), contribute to the development of an interface of analytical modelling and participatory approaches. Hereby, a potential resolution of the duality between individual and collective levels of a system is offered.

Agent based modelling is a rather new approach to simulate possible future land-use changes and resource management (GEBETSROITHER et al., 2004, 2006) at considering spatial behaviour of individual agents. This approach leads to very detailed, local results and thus provides the information needed by local authorities and inhabitants interested in information about future land-use change consequences for their own neighbourhoods. Application of ABM in the spatial sciences started in the mid-1990s (TORRENS, 2001). ABM models, initiated from sciences like psychology, sociology and biology, did not originally treat space explicitly. The application of ABM in spatial sciences has reshaped agents' activities from immobile decision making and communication to mobile action within a virtual space, like movement through a landscape or along roads modelled through a set of raster layers. Raster-cells arc applied here as entities whose state changes not (only) because of agent behaviour searching for a certain cell and changing the state of this underlying cell - based on certain behavioural rules. ABM approaches address single agent behaviour over time through iterative steps, executing destination search cycles of single agents, which perform steady land-use changes inside the study region. (LOIBL et al., 2003, 2006). Through land use change probability modelling (within this volume: LOIBL and GEBETSROITHER, 2007) attractiveness layers for future development have been derived related with a set of rules for different agent classes, where and how their actions will be performed.

The model framework was developed using REPAST – an Open Source Java-based toolkit to develop agent based models (http://repast.sourceforge.net). As REPAST was developed to model social science aspects without considering space and location explicitly as GIS does further functionalities had to be included.

3 Integration of GIS information, system dynamics and Agent Based Modelling

There are three main approaches of coupling environmental models with GIS described in (HUANG and JIANG, 2002), *loose coupling, tight coupling* and *full coupling*. For our purpose the full coupling was the appropriate approach. As a GIS is a tool delivering static information as a model's target result, an independent tool has been written in Java making use of the REPAST ABM-toolkit which has to be extended to handle spatial information and allows to perform space related action of agents that trigger land use and water demand allocation dynamically. The developed ABM-GIS link handles raster-cell layers using ESRI's ASCII grid-format which serves as a quasi-standard for GIS-raster-data exchange. The developed platform provides import/export, calculation of new raster-layers out of existing ones using simple arithmetic operations on the one hand and agent's spatial movement transforming land use of single cells leading to steady land use transition. The developed graphical user interface (GUI) provides basic GIS-like viewing functionality (until now: zooming, panning, layer selection, map-legends and cell value retrieving functionality) (Fig. 2).

The connection to the System Dynamic model (built with Vensim) was established via dynamic link libraries (DLL's). A Vensim interface was also integrated into the platform to steer the Vensim simulation and to visualise the Vensim specific results. This enables us for example to change constants and tables used as input in the Vensim model without directly working in Vensim (Fig. 3). Changes made during the simulation can be recorded automatically (this is a feature of Vensim) and therefore different simulation runs can be compared at a later stage.

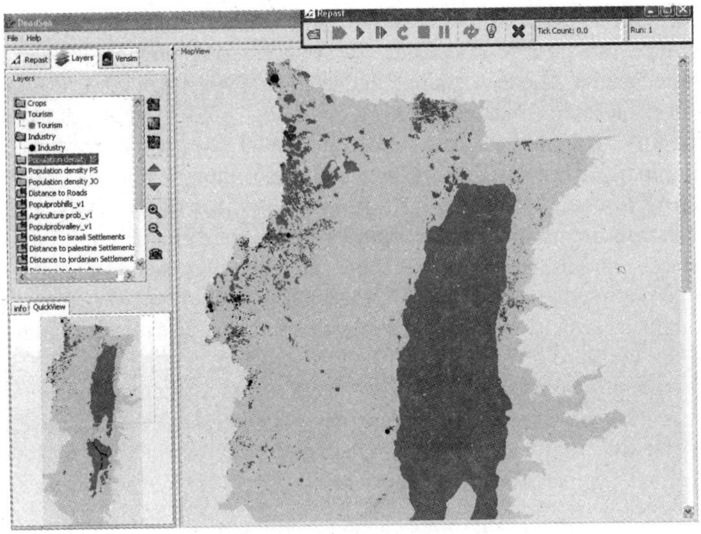

Fig. 2: GISmo Graphical User Interface (GUI)

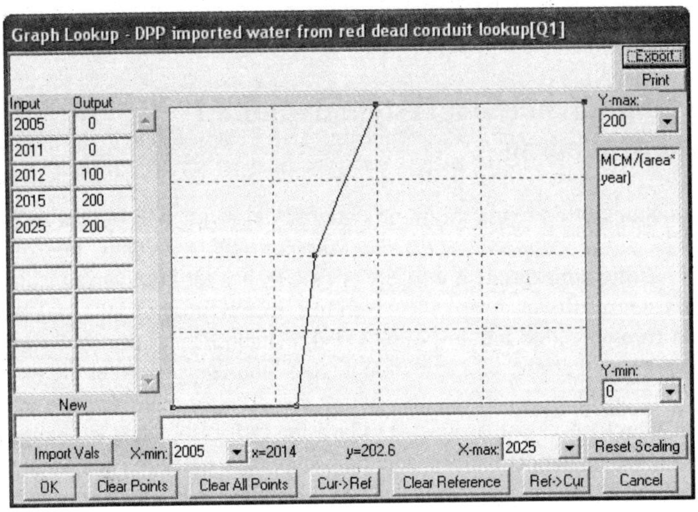

Fig. 3: Integrated Vensim table

4 Results and conclusion

Figure 4 shows as example land use changes between 2005 and 2025, due to the agents' activities. Regions with high activity changes can be extracted, used for planning purposes and as a starting point for discussion with local stakeholders. Because of the integration of the two approaches also top down information e.g., about the Dead Sea water level development (Fig. 5) is available for the decision makers. Further information about the project and integration platform can be found at http://maps.arij.org/deadseaproject/.

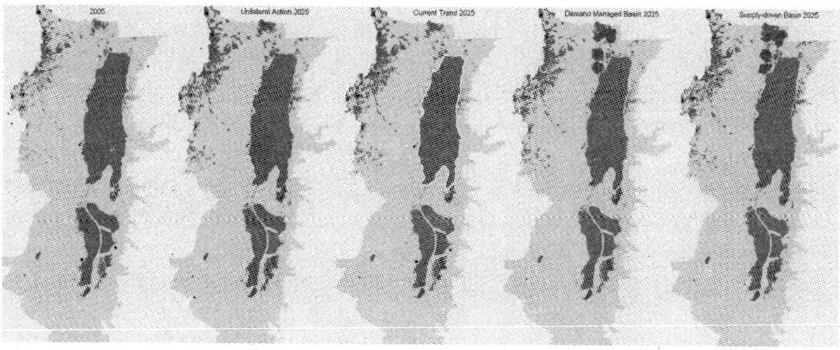

Fig. 4: GIS development maps 2005 left, scenario results 2025 right

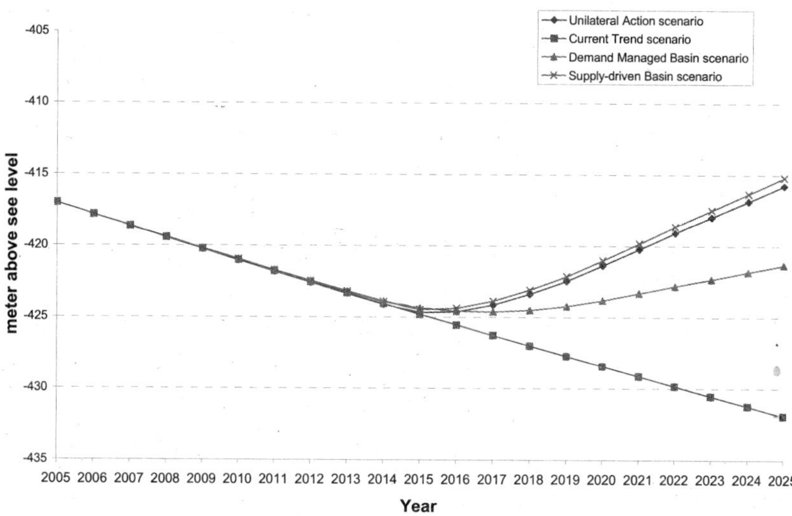

Fig. 5: Development of the Dead Sea level within four different scenarios

The integration of System Dynamics and Agent Based Modelling with a GIS enables to support decision makers with important information about the local, spatial, effects of their measures and the feedback reactions of the actors to these policies and enables them to find more sustainable options for the future.

References

BARRETEAU, O., GARIN, P., DUMONTIER, A., ABRAMI, G. and CERNESSON, F., 2003, Agent-based facilitation of water allocation: Case study in the Drome River Valley. Group Decision and Negotiation 12(5), pp. 441–461.

BECU, N., PEREZ, P., WALKER, A., BARRETEAU, O. and LE PAGE, C., 2003, Agent based simulation of a small catchment water management in northern Thailand description of the CATCHSCAPE model. Ecological Modelling 170(2–3), pp. 319–331.

BENENSON, I. and TORRENS, P.M., 2004, Geosimulation: Automata-based Modeling of Urban Phenomena. Wiley, Chichester.

BERGER, T., 2001, Agent-based spatial models applied to agriculture: a simulation tool for technology diffusion, resource use changes and policy analysis. Agricultural Economics 25(2–3), pp. 245–260.

CLARKE, K.C., HOPPEN, S. and GAYDOS, L.J., 1997, A self modifying cellular automata model of historical urbanization in the San Francisco Bay area, Environment and Planning B. Planning and Design, 24(2), pp. 247–261.

DRAY, A., PEREZ, P., JONES, N., LE PAGE, C., D'AQUINO, P. and AUATABU, T., 2006, The Atoll Game Experience: from Knowledge Engineering to a Computer-Assisted Role Playing Game. Journal of Artificial Societies and Social Simulation, 9(1).

GEBETSROITHER, E., KAUFMANN, A., GIGLER, U. and RESETARITS, A., 2004, Modelling self-organization processes in socio-economic and ecological systems to support adaptive forest management. Modelling Forest Production, Vienna, April 19–22nd 2004.

GEBETSROITHER, E., KAUFMANN, A., GIGLER, U. and RESETARITS, A., 2006, Agent-Based Modelling of Self-Organisation Processes to Support Adaptive Forest Management. In BILLARI, F.C., FENT, T., PRSKAWETZ, A. and SCHEFFRAN, J. (Eds.), Agent-Based Computational Modelling. Physica, Heidelberg, pp. 153–172.

GEBETSROITHER, E. and STREBL, F., 2003, Coupling of Geographic Information System and System Dynamic Models (A regional spatio-temporal soil erosion model implemented with ArcMap and Vensim). In Proceedings of the Fourth International Symposium on Mathematical Modelling, February 5–7, 2003, Vienna University of Technology. ARGESIM-Reports ISBN 3-901608-24-9.

GIDDENS, A., 1979, Central Problems in Social Theory: Action, Structure, and the Contradiction in Social Analysis. Berkeley: University of California Press.

GILBERT, N. and CONTE, R., 1995, Artificial societies: the computer simulation of social life. London, UCL Press Limited.

LOIBL, W. et al., 2006, Simulation of polycentric urban growth dynamics through agents – Model concept, application, results and validation. In KOOMEN, E., STILLWELL, J., BAKEMA, A. and SCHOLTEN, H.J. (Eds.), Modelling Land-Use Change. Springer Netherlands, pp. 219–235.

LOIBL, W. and GEBETSROITHER, E., 2007, Middle East-Water Challenge – Supporting Decision and Negotiation Processes through Modelling Land Use Change Probability in the Dead Sea Area, Considering Alternative Water Supply Policies. In ZEIL, P. and KIENBERGER, S. (Eds.), Geoinformation for Development. Wichmann, Heidelberg, pp. 147–153.

LOIBL, W. and TÖTZER, T., 2003, Modelling Growth and Densification Processes in Sub-urban Regions – Simulation of Landscape Transition with Spatial Agents. Environmental Modelling and Software, 18(6), pp. 485–593.

STERMAN, J., 2000, Business Dynamics, Systems Thinking and Modelling for a Complex World. McGraw-Hill Companies, p. 982 ff.

TILLMAN, T., LARSEN, T.A., PAHL-WOSTL, C. and GUJER, W., 2001, Interaction analysis of stakeholders in water supply systems. Water Science and Technology, 43(5), pp. 319–326.

TORRENS, P.M., 2001, Can Geocomputation Save Urban Simulation? Throw some agents into the mixer, simmer and wait ... . CASA Working Paper Series, 32. Centre for Advanced Spatial Analysis, University College London. http://www.casa.ucl.ac.uk/working_papers/paper32.pdf

Probability Modeling to Reduce Decision Uncertainty in Environmental Niche Identification and Driving Factor Analysis – CaNaSTA Case Studies

Karl ATZMANSTORFER, Thomas OBERTHÜR, Peter LAEDERACH, Rachel O'BRIEN, Laura COLLET and Gettsy Elisabeth QUIÑONEZ

Summary

Hillside agro-ecosystems have a complex spatial and temporal distribution of natural resources. Farmers generally possess a vast body of knowledge about environmental resources on their farms but this knowledge is largely based on locally observable features rather than generalized knowledge. The lack of process-based knowledge concerning agro-ecosystem function creates uncertainty that obstructs sound decision-making under conditions of rising economic and ecologic pressure in many developing countries. Since the past decade, Precision Agriculture provides tools to reduce uncertainty caused by environmental variation. By describing spatial and temporal variation of the environment, Geographic Information Systems help to detect suitable crops for specific environmental niches and support farmers to find optimal management practices for their plot of land. Hence Precision Agriculture helps to raise the economic benefits of farming, ensures consistent product quality and reduces negative environmental impacts caused by inappropriate management practices.

A spatial decision support system called CaNaSTA was developed to aid the decision making process of crop adoption in tropical agriculture. Using Bayesian probability statistics, CaNaSTA integrates trial data, spatial data and expert knowledge and provides maps, tables and graphs analyzing and interpreting the probability distributions of spatial phenomena. The International Centre for Tropical Agriculture (CIAT) has applied CaNaSTA to three case studies related to tropical agriculture. The first case study identifies niches for specialty coffee production, the second analyses the potential of cowpea (*Vigna unguiculata (L.) Walp.*) for tropical hillside environments in Colombia. Finally, Canasta was applied to a non-crop related area by performing a study of carbon concentration in tropical soils.

1 Introduction

A complex spatial and temporal distribution of natural resources is characteristic for hillside agro-ecosystems. Site-specific management further compounds the naturally introduced, uncontrollable spatial and temporal variation. Considerable evidence exists that farmers identify and use variation at scales relevant for management. Encouraging as this is, increasing land degradation and prevailing poverty in developing countries suggest that locally devised methods, on their own, are no longer effective enough to cope with rapidly changing pressures. Farmers generally possess a vast body of knowledge about environmental resources on their farms but this knowledge is largely based on observable features rather than generalized knowledge. The lack of process-based knowledge concerning agro-ecosystem function and its temporal or spatial variability creates uncertainty that obstructs sound decision-making under conditions of change and provokes farmers to generally underestimate existing environmental variability (COOK, 1999; COOK and BRAMLEY, 2000). Hence this uncertainty prevents farmers from taking decisions that are too risky and may have contributed to their reputation of being risk-averse.

Over the past decade, Precision Agriculture has provided tools to reduce uncertainty caused by environmental variation, in order to improve crop management (DIXON and MCCAIN, 1997; COOK and BRAMLEY, 2000). Site-specific management practices – as an integrative part of Precision Agriculture – use information technologies like GPS, GIS, Remote Sensing and computer models tailoring crop management to fit specific environmental conditions (LOWENBERG-DEBOER, 2000). By describing spatial and temporal variation of the environment, these techniques help to detect environmental niches and support farmers to find optimal management practices for their plot of land (ROBERT, 1996). They enable farming to be analyzed as controllable systems, rather than a disaggregated set of agronomic and environmental factors and therefore help to better estimate the influence of environmental factors like climate, soil or topography on crop production or product quality (COOK et al., 2004). Hence site-specific management practices help to raise the economic benefits of farming, ensure consistent product quality and reduce negative environmental impacts caused by inappropriate management practices (BLACKMORE, 2002; COOK and BRAMLEY, 2000).

In the 1990s, a number of models for predicting crop response to climate, water, light and other conditions (SADLER and RUSSEL, 1997) or for predicting the geographic distribution of species in the wild (JARVIS et al., 2002; JONES and GLADKOV, 1999) have been developed. However, only a few of them included a spatial component precise enough for site-specific management practices, which in general require numerous data points. Hence the statistical approaches used in these models are not appropriate for many high value crops, which are not widely cultivated due to a lack of information.

2 Tools and methods

2.1 CaNaSTA

Especially in heterogeneous hillsides where many cultivars can vary within a very small area (OBERTHUER et al., 2004) it is of interest to predict which crop or variety can produce well in a given niche. Lately, researchers at CIAT have initiated the development of the "Crop Niche Selection for Tropical Agriculture" tool "CaNaSTA". CaNaSTA combines expert and local knowledge on the requirements of given crops or varieties, with the identification and characterization of a small number of specific sites where a crop or variety is known to grow well. CaNaSTA uses this information in order to identify other similar niches where a crop or variety is likely to perform in the same way (OBERTHUER et al., 2004). By making the spatial variability of environmental factors explicit, CaNaSTA supports farmers in order to make less risky agronomic decisions even when there is just limited data available, helping to rapidly identify potential production areas for agricultural products saving time and money.

The main difficulty in modelling environmental variability is related to the attempt to represent complex, spatially and temporally variable behaviour with simplified models, which use sparse input data (COOK and ADAMS, 1999). CaNaSTA was particularly designed for such situations, when only sparse data is available or data generation is too expensive. The tool operates on the basis that the state of a particular property, which may be difficult to measure directly over a wide area (e.g. soil carbon content or the incidence of speciality coffee), can be inferred from the state of other more easily measurable entities (e.g. climate or topographic data) and the knowledge of their interrelationships. CaNaSTA uses probability theory (Bayesian statistics) to combine these disparate data sets and provides maps, tables and graphs in order to visualize and interpret the probability distributions of spatial phenomena at specified levels (O'BRIEN et al., 2004).

2.2 Environmental factor generation

As a first step of the CaNaSTA model set-up, decisive environmental variables have to be identified and mapped in order to calculate probability surfaces for the model's response variable (which can be e.g. sensorial coffee quality, crop yield or soil carbon content). For the case studies mentioned in Chapter 3.1–3.3, we selected up to 14 different environmental factors representing environmental variability, mapped and integrated them to CaNaSTA as the model's predictor variables. Resolution for the factor maps is 30 arc and 3 arc seconds, which correspond in Colombia to a 1 km and a 90 m grid size, respectively (O'BRIEN, 2004).

For generating data of *climatic factors* in the tropics, it has to be often resorted on interpolated climate surfaces, as generally data of only a few weather stations is available in these regions (JONES and THORNTON, 2000). In our model we extracted and mapped climate data using WorldClim data, which is a set of global climate layers at a resolution of 30 arc seconds. The layers were generated through interpolation of average monthly climate data, collected from up to 47 000 weather stations of a period from 1960 to 1990 (HIJMANS et al., 2005). We extracted and generated the following climate factors using WorldClim data: annual precipitation, annual temperature, dry months per year, diurnal temperature range, dew point and solar radiation. We defined dry months as months with less than 90 mm of precipitation. We calculated mean diurnal temperature range according to JONES et al. (2002) as the difference between mean monthly maximum and mean monthly minimum temperature. We mapped dew points by the method of LINACRE (1977) and calculated mean annual solar radiation according to the method of KUMAR et al. (1997) using digital elevation and latitude data.

Identifying and preparing *pedologic factors* for CaNaSTA use was the most challenging task during the data preparation process as it was not possible to map specific soil attributes, which heavily influence the models' response variables. Soil characteristics vary widely at a small scale in the study area, but unfortunately there are no soil maps at a high enough resolution to adequately represent the soil heterogeneity available for Colombia. Although various general characteristics such as pH-value and fertility can be derived from maps representing soil types, direct correspondence between soil type and other soil characteristics is less certain (O'BRIEN, 2004). Therefore, we digitized the 1:500 000 soil map of the Agustín Codazzi Geographic Institute (IGAC, 1981) and simply used the corresponding soil types for the analysis. We generated and mapped *topographic factors* (elevation, aspect and slope) using a processed DEM of the Shuttle Radar Topography Mission (SRTM).

3 Applications in CIAT – case studies

CIAT applied CaNaSTA to three case studies related to tropical agriculture, which are presented in the following chapter. The first case study identifies niches for high quality coffee production, the second study analyses the potential of cowpea (*Vigna Unguiculata (L.) Walp.*) as component of animal feed concentrates. Both analyses were performed for tropical hillside environments in the Colombian Province of Cauca. Finally, CaNaSTa was applied to a non-crop related area by performing a study of carbon concentration in the tropical soils of Fúquene Lagoon Basin in Cundinamarca Province, Colombia.

3.1 Coffee case study

Around 25 million families in hillside regions of the tropics depend on cultivation of coffee as the major source of their income. Steadily declining and highly volatile coffee prices are jeopardizing the livelihood of thousands of smallholder farmers for nearly 40 years now. From 1977 to 2001 real international prices for coffee fell by 5.1% yearly; in 2002 the price of coffee was at a 30 year low – 25% less of that in 1960 (GRESSER and TICKELL, 2002). This alarming development and its related socio-economic and ecological consequences (FLORES et al., 2002; VARANGIS et al., 2003) have shifted the attention of coffee-related research from increasing production volumes to specialty coffee. Within the framework of the Diversification Agriculture Project Alliance (DAPA), CIAT is investigating promising economic alternatives for diversifying smallholder coffee agro-ecosystems. In this context specialty coffee was identified as one promising economic alternative to commodity coffee, besides other cultivars like tropical fruits, honey, medicinal plants and cowpea.

3.1.1 Materials

Sensorial coffee quality depends on environmental and genetic factors, as well as on agronomic management of the crop and post-harvest practices. In order to analyze the barely investigated dependency of coffee cup quality on environmental factors, we carried out a case study linking coffee quality data from 88 farms to the ten climatic, topographic and pedologic factors, mentioned in Chapter 2.2. Study area was the Province of Cauca in Southern Colombia. Cauca is a rural department with agricultural production as the main source of income. More than 60 000 ha coffee was cultivated in 2002 by more than 90 000 mainly small and medium producers (COLLET, 2005).

We used data from literature and expert consultations as indicators in order to identify sample farms with potential to produce speciality coffee. Within each farm, the coffee samples were harvested on Management Unit (MU) base. We defined a MU as one or several fields with homogenous environmental and agricultural management characteristics, with a minimum of about 5000 coffee trees. We geo-referenced each MU with latitude, longitude, and elevation in the centre of each MU by using GPS measurements. 90 percent of the samples were harvested from the variety *Caturra*, the remaining ten percent from the varieties *Colombia* and *Bourbon*. Samples were subject to standardized post-harvest management, in order to assure comparable results. All the samples of the study were cupped by a panel of five cuppers according to international cupping standards including e.g. aroma, flavour, after-taste, acidity and uniformity of the coffee.

3.1.2 Results

We carried out different analyses using CaNaSTA for relating environmental conditions to cup coffee quality (only selected results are presented here). In order to delimitate the study area to zones with specific quality characteristics, we used CaNaSTA for estimating the most probable quality per niche (Fig. 1). This probability surface shows that large area produce mediocre coffee but only very limited areas can produce superior quality coffee. Nevertheless, we can identify ecological niches with high probabilities of producing specialty coffee using CaNaSTA. The niches identified are situated in the municipalities of Tambo and Timbío in the Popayán area and the municipality of Inzá on the eastern border of Cauca Province. These municipalities are commonly known as areas where specialty coffee is produced.

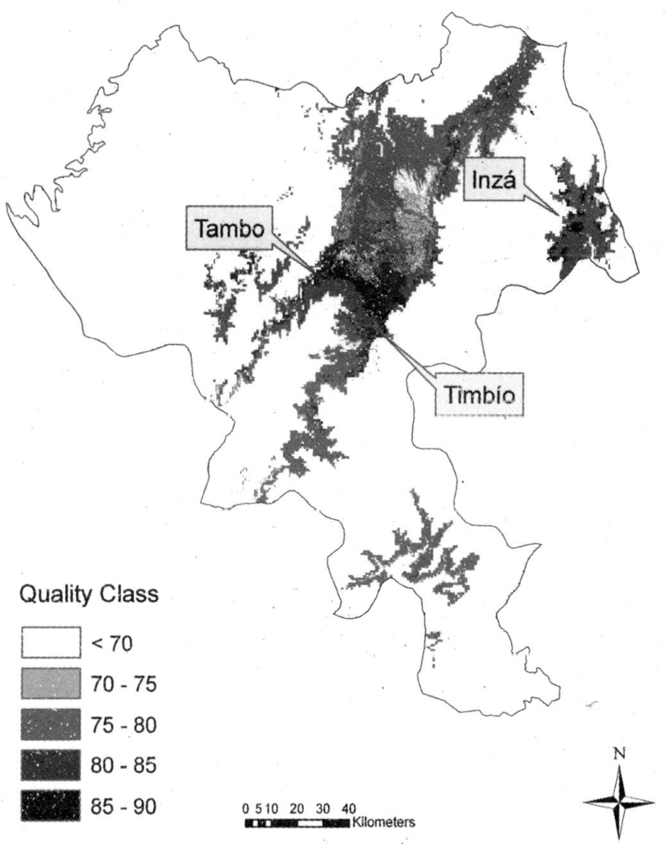

Fig. 1: Most probable coffee quality class, Cauca Province

The localization of niches in the same areas can be observed when we generate maps for individual coffee quality characteristics such as acidity, sweetness, and body.

Besides identifying environmental niches that are able to carry high quality coffee, it is also of interest to know which environmental factors are prone to improve or derogate coffee quality. When analysing the factors that most affect cup quality ("driving factors") with CaNaSTA, it became obvious that especially the *combination* of specific factors drive coffee quality. The impact of elevation on coffee quality, for example, seems to depend on the aspect of a slope where coffee is grown. Not only specific factor combinations, but also specific factor *ranges* seem to be important for coffee quality. CaNaSTA showed that altitudes from 1800–2000 m in the study area have a positive impact on quality, whereas other altitude categories have a negative or not a significant impact. Regarding annual median temperature as a driving factor for coffee quality, a range of 16–18 °C show a positive influence, while a range of 19–20 °C show a negative one.

3.2 Cowpea case study

As argued in the coffee case study above, one promising alternative for diversifying commodity coffee cultivations is speciality coffee, commonly grown at altitudes above 1500 m. Coffee cultivated in so called "marginal coffee growing areas" at lower altitudes between 1000–1500 m is usually of marginal quality and therefore relies on the highly volatile markets for volume coffees (GRESSSER and TICKELL, 2002). Hence alternative crops should be found for cultivation in these areas. CIAT identified forage seed and leaf meals produced from cowpea (*Vigna Unguiculata* (L.) *Walp.*) as a possible way for many smallholders in these regions to substitute commodity coffee cultivation.

Cowpea, one of the most widely grown legumes in tropical and subtropical regions, is of major importance to the livelihoods of millions of poor people, especially in Africa and China (PEREZ-LOPEZ et al., 2003). Cowpea provides inexpensive and nutritious food (grain, pods), gives excellent forage (grain concentrate, leaves and haulm), hay and silage. The plant improves soil quality by fixing soil nitrogen. It is drought and heat resistant, and as a fast-growing ground-cover plant it prevents erosion (QUIN, 1997; COOK et al., 2005). At present, cowpea is not widely grown in most of Colombia or other tropical hillsides in Latin America and the Caribbean (LAC). Very scarce evidence exists of its use as an input to commercial feeds and its production potential. Moreover its performance in the hillside eco-region is not yet fully understood.

Therefore the general objectives of this case study were (a) to identify optimal growing areas for cowpea by examining the effects of environmental variability on grain and biomass production using CaNaSTA, and (b) to estimate financial viability of cowpea as component of animal feed concentrates cultivated by smallholders in remote areas as substitute crop for locally traded soy.

3.2.1 Materials and methods

We established four different cowpea accessions (DICTA 9611, IITA 5234, IITA 1088-4 and IITA 1069-6) at seven different sites located in contrasting landscape positions in the study area, the Cauca Province. Trial sites were 2 100 m^2 each – areas big enough to give farm-size agronomic and therefore economically valid data. We measured coordinates and areas of each trial site with GPS and integrated grain and biomass yield data as point data to the CaNaSTA model. We identified more yield data at sites in the Colombian Provinces of Valle de Cauca and Nariño as well as in Honduras revising the literature and added it to the model as one way of integrating expert knowledge to CaNaSTA in a data sparse situation.

In order to represent environmental variation in the study area, we mapped 14 different environmental factors. In addition to the climatic, topographic and pedologic factors used in the coffee case study, we derived four hydrologic indices from the DEM. These indices are considered as good indirect measures for the spatial variability of hydrological processes (MOORE et al., 1991), which play an important role in the local composition of soils. For modeling soil water content and surface saturation zones we derived a "Topographic Wetness Index", TWI (MOORE et al., 1991) and a "Slope-length factor" (MOORE and WILSON, 1992) using Arc View 3.2 "Topocrop Terrain Indices" Extension (SCHMIDT, 2003). Furthermore, we used "plan curvature" and "profile curvature" for describing the spatial pattern of the deposition of nutrients and other soil components in our model (ZEVERBERGEN et al., 1997).

3.2.2 Results

In (a) the *spatial analysis* we examined the adaptation of the above-mentioned cowpea accessions to specific environmental conditions analysing grain and biomass yield. We used CaNaSTA to identify areas in marginal coffee growing regions of Cauca Province between 1000–1500 m where cowpea is likely to adapt well. This data was required to estimate the potential area for cowpea production and hence the potential quantities available as animal feed concentrate. Figure 2 shows the probability of cowpea adaptation – dark areas show high probabilities of adaptation (up to 75%), light areas low probabilities.

CaNaSTA also mapped the *certainty* of the prediction based on the number of trial data entries, the *stability* of cowpea adaptation in case of changing environmental conditions and the specific influence of each of the environmental factors used in the model on cowpea adaptation (results not presented). We also determined the best cowpea accession for selected locations (farms) using CaNaSTA. *Driving factor analysis* revealed several ranges of environmental factors driving cowpea adaptation. As positive driving factor ranges CaNaSTA identified warm temperatures (24.1–27.5 °C), low altitudes (0–700 m), a low diurnal temperature range (6.8–9 °C), a high dew point (18.1–23.4 °C) and a high TWI (9-15; that means areas, highly saturated with water). Negative driving factor ranges are slightly flat to medium slopes (6–10 °) and slopes with northern aspect (NW–NE).

CaNaSTA enabled us to calculate the most likely cowpea yield category occurring at each cell of the study area. We used this information in (b) the *financial analysis* for deriving income surfaces, showing possible revenues of cowpea cultivation at an estimated cowpea price for each cell. Revenues up to nearly 2 400 000 Colombian Pesos (approximately 1 050 US$) could be achieved per hectare in the best performing niches. For deriving this map, we examined production costs at each trial site in order to determine yield-price-profit thresholds. CaNaSTA results showed that cowpea could be an alternative crop for diversifying marginal coffee regions in specific environmental niches.

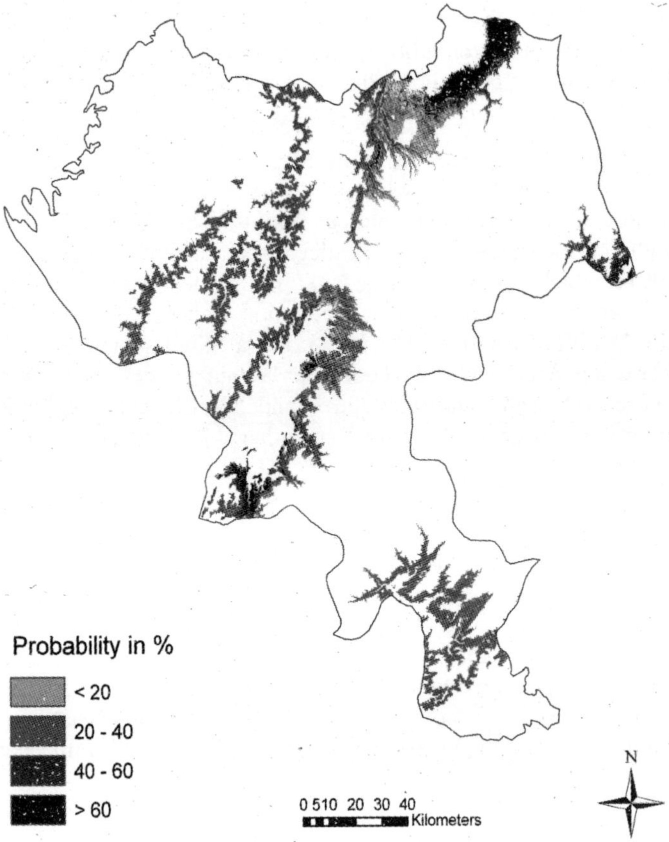

Fig. 2: Probability of cowpea adaptation, marginal coffee growing areas, Cauca Province

3.3 Fúquene Lagoon Basin case study

In a third case study, researchers of CIAT are using CaNaSTA for deriving probability surfaces of carbon concentration in the soils of Fúquene Lagoon Basin in the Colombian Province of Cundinamarca. In this area the decrease of natural vegetation cover due to intensive agricultural use over the past decades resulted in a constant loss of organic matter, causing diminishing carbon contents in the Lagoon Basin's soils (RONDON and AMEZQUITA, 2005). The main aim of the investigation is to analyze under which environmental conditions, and in which soil horizons a maximum of carbon is absorbed. This case study should give an example, how CaNaSTA can be applied also on non-crop related topics in the near future. Preliminary results will be available in autumn 2006.

4 Discussion and conclusion

The CaNaSTA methodology used in the above-mentioned case studies provides insight into the interaction of agronomic and ecologic variables (e.g. sensorial coffee quality, cowpea yield, carbon soil concentration) with environmental conditions that was not previously available. Valuable results can be achieved by calculating probability surfaces integrating trial data and expert knowledge to a GIS-based model instead of exclusively using time- and resource-consuming field trials. CaNaSTA proved to be an effective modelling tool in data sparse situations when there are not enough data available for common statistical analyses. Using local and expert knowledge, CaNaSTA approximated coffee quality and cowpea yield for very small-scale areas (90 m grid-resolution) with highly varying environmental conditions. The results proved that CaNaSTA can provide farmers with valuable information for choosing the right cultivar for a specific environmental niche or for choosing appropriate management practices in order to reduce agronomic decision uncertainty rooted in temporal and spatial variability of the surrounding environment.

The most limiting factor for CaNaSTA application is lacking environmental data at high-resolution. In particular, soil data is not commonly available at a resolution high enough to represent its spatial variability. In the near future CIAT will focus on the use of CaNaSTA as a spatial analysis tool in order to delimit environmental niches where specialty coffee has a high probability to grow well. The approach used in the case studies will be applied for the denomination of coffee origin by identifying environments that are capable of producing coffee of similar quality profiles. CaNaSTA will also allow definition of zones that are less appropriate for coffee production where coffee should be replaced with other crops like cowpea, tropical fruits, honey or medicinal plants in order to ensure sustainable farmer income. In the long term, CaNaSTA should provide means for quickly identifying potential production areas for agricultural products, thereby helping farmers to engage in profitable supply chains on national and international agricultural markets.

References

BLACKMORE, S., 2002, Developing the principles of precision farming. The Center of Precision Farming, The Royal Veterinary and Agricultural University, Denmark.

COLLET, L., 2005, Impact of Environmental Factors to Manage Coffee Cup Quality. Université Paris 12 Val de Marne. Créteil, France.

COOK, S. and ADAMS, M.L., 1999, Uncertainty and Interpretation of Spatial Information. Wembley, Australia: CSIRO Land and Water Department.

COOK, S. and BRAMLEY, R., 2000, Precision Agriculture: Using paddock information to make cropping systems internationally competitive. Conference Proceedings, Emerging Technologies in Agriculture Conference. Sydney, Australia.

COOK, S., O'BRIEN, R., OBERTHÜR, T. and RUBIANO, J., 2004, Is precision agriculture irrelevant to developing countries? Conference Proceedings, 7th International Precision Agriculture Conference. Minneapolis, USA.

COOK, B.G., PENGELLY, B.C., BROWN, S.D., DONELLY, J.L., EAGLES, D.A., FRANCO, M.A., HANSON, J., MULLEN, B.F., PATRTRIDGE, I.J., PETERS, M. and SCHULTZE-KRAFT, R., 2005, Tropical Forages: an interactive selection tool. CD-ROM. Brisbane, Australia: CSIRO, DPI&F, CIAT and ILRI.

DIXON, J. and MCCAIN, M., 1997, Precision Agriculture in the 21st Century. Washington D.C., USA.

FLORES M., BRATESCU, A., MARTINEZ, J.O., OVIEDO, J.A. and ACOSTA, A., 2002, Centro-américa: El impacto de la caida de los precios del cafe. NY, USA: United Nations.

GRESSER, C. and TICKELL, 2002, Mugged – Poverty in your cup. Oxfam International. Online Report. http://www.maketradefair.com/assets/english/mugged.pdf.

HIJMANNS, R.J., CAMERON, S.E., PARRA, J.L., JONES, P.G. and JARVIS, A., 2005, Very high resolution interpolated climate surfaces for global land areas. International Journal of Climatology, 25, pp. 1965–1978.

IGAC (Augustin Codazzi Geographic Institute), 1981, República de Colombia: Mapa de suelos. Escala 1:1 500 000. Bogotá, DE, Colombia.

JARVIS, A., GUARINO, L., WILLIAMS, D., WILLIAMS, K. and HYMANN, G., 2002, Spatial analysis of wild peanut distributions and the implications for plant genetic resource conservation. Plant Genetic Research Newsletter, 131, pp. 29–35.

JONES, P. and GLADKOV, A., 1999, FloraMap: A computer tool for the distribution of plants and other organisms in the wild. Cali, Colombia: CIAT.

JONES, P. and THORNTON, P., 2000, MarkSim – Software to Generate Daily Weather Data for Latin America and Africa. Agronomy Journal, 92, pp. 445–453.

JONES, P., THORNTON, P., DIAZ, W. and WILKENS, P., 2002, MarkSim – A computer tool that generates simulated weather data for crop modeling and risk assessment. CD-ROM. Cali, Colombia: CIAT.

KUMAR, L., SKIDMORE, A.K. and KNOWLES, E., 1997, Modeling topographic variation in solar radiation in a GIS environment. International Journal for Geographical Information Science, 1(5), pp. 475–497.

LINACRE, E., 1977, A simple formula for estimating evaporation rates in various climates, using temperature data alone. Journal of Agricultural Meteorology, 18, pp. 409–424.

LOWENBERG-DEBOER, J., 2000, Precision Farming Profitability. Site-Specific Management Center, Purdue University, West Lafayette, Indiana, USA.

MAIZELS, A., 1994, The Continuing Commodity Crisis of Developing Countries. World Development Journal, 22(11), pp. 1685–1695.

MOORE, I.D., GRAYSON, R.B. and LADSON, A.R., 1991, Digital Terrain Modeling: A Review Of. Hydrological, Geomorphologic and Biological Applications. Hydrological Processes, 5, pp. 3–30.

MOORE, I.D. and WILSON, J., 1992, Length-slope factors for the Revised Universal Soil Loss Equitation: Simplified method of estimation. Journal of Soil and Water Conservation, 47(5), pp. 423–428.

OBERTHÜR, T., COOK, S., LUNDY, M., WATTS, G., O'BRIEN, R. and JONES, P., 2004, Research Framework that support sustainable diversification of coffee-based cropping systems with higher value crops. CIAT, Cali, Colombia.

O'BRIEN, R., 2004, Spatial decision support for selecting tropical crops and forages in uncertain environments. Thesis Doct. Philosophy. Curtin University of Technology, Australia.

O'BRIEN, R., PETERS, M., CORNER, R. and COOK, S., 2004, CaNaSTA – Crop Niche Selection for Tropical Agriculture, a spatial decision support system. Annual Report 2004, Project IP-5. CIAT, Cali, Colombia.

PEREZ-LOPEZ, O., PEREZ-BONNA, R.A., GUZMAN, G. and MOJICA-RODRUIGEZ, J.E., 2003, Cultivos para alimentación animal en sistemas de producción bovina de la Orinoquia colombiana. Bogotá, Colombia: Corpoica, Ministerio de Agricultura y Desarrollo Rural de Colombia and CIAT.

QUIN, F.M., 1997, Introduction. In SINGH, B.B, MOHAN RAJ, D.R., DASHIELL, K.E. and JACKAI, L.E.N. (Eds.), Advances in Cowpea Research. Co-Publication of International Institute of Tropical Agriculture (IITA) and Japan International Research Center for Agricultural Sciences (JIRCAS). Ibadan, Nigeria, pp. ix–xv.

ROBERT, P.C., 1996, Precision Agriculture. Proceedings of the 5th Precision Agriculture Conference on Precision Agriculture, Precision Agriculture Center, University of Minnesota, USA.

RONDON, M. and AMEZQUITA, E., 2005, Propuesta para la realización de inventarios de Carbonos y medición de Gases de efecto invernadero en los sistemas de uso dominante en la Cuenca de la Laguna de Fúquene. Unpublished Document. CIAT, Cali, Colombia.

SADLER, E.J. and RUSSELL, G., 1997, Modeling crop yield for site-specific management. In ROBERT, P.C., RUST, R.H. and LARSON, W.E. (Eds.), The State of Site-Specific Management for Agricultural Systems. Madison, USA, pp. 69–80.

SCHMIDT, F., 2003, Hochgenaue Digitale Gelaendemodelle – Untersuchungen zur Erstellung, Analyse und Anwendung in der Landwirtschaft. Universät Rostock, Germany.

VARANGIS, P., SIEGEL, P., GIOVANNUCCI, D. and LEWIN, B., 2003, Dealing with the Coffee Crisis in Central America: Impacts and Strategies. The World Bank, Washington D.C., USA.

ZEVERBERGEN, L.W. and THORNTON, C.R., 1997, Quantitative Analysis of Land Surface Topography. Earth Surface Processes and Landforms, 12, pp. 47–56.

Capacity Development Within the BIOTA East Africa Project – Promoting the Use of Spatial Information in Biodiversity Research and Management

Gertrud SCHAAB

Summary

Within the BIOTA East Africa project, funded by the German Federal Ministry of Education and Research (BMBF) since 2001, the influence of fragmentation and human use on the biodiversity of East-African rainforests is investigated following an interdisciplinary and integrated research approach. Besides providing time series of land use/cover change and intra-annual variations in biophysical parameters to the biologists, ecologists, agronomists and economists involved, subproject E02 aims at developing capacities in the use of spatial information. Here, our project partners with a biological background have been surprisingly unaware of the potential resulting from data analysis and processing including a spatial reference. We therefore consider interdisciplinary cooperation as part of capacity building, which also includes specific training of young scientists from the counterpart institutions in Kenya and Uganda. As the consortium has promised to come up with landscape level recommendations for a sustainable use and conservation of East-African rainforests, biodiversity researchers are not only in need of geo-spatial data but should work on spatially explicit representations of their findings. Also sound exit strategies are required for terminating the project, i.e. the funding. By that time the East-African counterparts should be enabled to continue with monitoring, analysing and deciding upon measures without further close support. The paper highlights the specific contributions of subproject E02 in regard to promoting the use of spatial information in biodiversity research and management.

1 Introduction

Biodiversity research is in need of geo-spatial data, especially when working at the landscape level and including global change aspects. Therefore, geo-spatial data need not only describe the recent environment but also changes that have occurred in the past or are likely to happen in the future. Nowadays GIS and remote sensing are widely used techniques whenever spatial patterns, temporal changes and functional relationships are to be analysed. All three aspects are of importance when aiming at a far-sighted conservation of biological diversity as well as at sound management concepts for sustainable use and fair benefit sharing. As described in SCHAAB et al. (2004) GIS and remote sensing can play a major role in support of goals and agreements of the Convention on Biological Diversity (CBD, 2000; www.biodiv.org/convention/articles.asp). Beside the general capabilities in CBD support, it has been shown in various studies over the last years that GIS and remote sensing are indeed useful for biodiversity research and monitoring (e.g. BERENS et al., subm.; FULLER et al., 1998; GOTTSCHALK, 2002; PETERS et al., subm.; ROY and TOMAR,

2000; SCHAAB et al., 2004). Further, it has been pointed out that a landscape's history can contribute to the understanding of the role of past natural and human disturbances, and thus provides the basis for a stronger likelihood of biology conservation success. As FOSTER (2006) concludes, such a historical narrative is a very effective means for engaging and educating managers and the public. When combined with intensive studies of species and habitats, history provides a rich source of important information. It can be even more efficiently accessed and analysed by adding a spatial reference and converging the information in a GIS (see MITCHELL and SCHAAB, 2006).

1.1 The BIOTA East Africa project

BIOTA East Africa is a project funded by BMBF, the German Ministry of Education and Research since 2001 (see www.biota-africa.de or KÖHLER, 2004). Fifteen subprojects are investigating the influence of fragmentation and human use on the biodiversity of East-African rainforests following an interdisciplinary and integrated research approach. The project encompasses a strong scientific capacity building component for the East-African counterparts as it is aiming at a sound handing-over towards the end of the funding period. Research is related to the vegetation structure, including factors responsible for regeneration, to certain animal groups (emphasizing invertebrates), and to ecological interactions (such as seed dispersal or pollination). Since the start of the second project phase also socio-economic issues are included. In the first project phase it has become evident, that it is not just abiotic factors such as soil or climate that effect biological diversity but also forest fragmentation and in particular forest disturbance as a consequence of intensive human forest use. Ultimately, it is to the landscape level that recommendations for a sustainable forest and biodiversity management are aimed. The focus area of research within BIOTA East Africa is the Kakamega Forest in Western Kenya. Budongo Forest and Mabira Forest in Uganda have been selected in the second project to enable comparison across disturbance gradients both within and across separate forests (see SCHAAB et al., 2005).

1.2 Objectives of subproject E02

The setting-up of a central BIOTA-East-GIS has gained increasing importance with the progression of the project due to the planned recommendations for sustainable forest and biodiversity management at the landscape level. Within the interdisciplinary approach of BIOTA, the GIS serves as the common platform a) to link observations of the different subprojects via spatial information stored on natural and socio-economic variables, and b) to extrapolate or scale-up the multidisciplinary findings based on field work in space and time (SCHAAB et al., 2004). Remote sensing is the means for providing the needed time series of land cover change (e.g. LUNG, 2004; MITCHELL et al., 2006) as well as of intra-annual variations in biophysical parameters (like the leaf area index; MÜLLER et al., 2004). These time series are further assessed regarding biodiversity-related indicators (see e.g. LUNG and SCHAAB, in print) aiming towards spatial as well as temporal extrapolations of field findings of the project partners. Subproject E02 also aims at developing capacities in the use of spatial information, this instead of just providing time series to the biologists, ecologists, agronomists and economists involved. Our project partners with a biological background have been surprisingly unaware of the potential resulting from data analysis and processing including a spatial reference. But as the consortium has promised to come up with recommendations at the landscape level, biodiversity research is not only in need of geo-spatial data but should work on spatially explicit representations of their findings. This explains the central position of subproject E02 within the research framework.

2 Capacity development

2.1 Capacity building within BIOTA-East

Apart from the scientific results, BIOTA East Africa is in particular known for its strong capacity building component. This includes staff training which in the case of field assistants, also means employment for people living next to the forests. Support covers the financing of technical equipment and the setting-up of databases for local institutions. Assistance regarding educational facilities is given. Here, a nature trail, a tree nursery, and a pollinator garden have been implemented in Kakamega Forest. Maseno University close by has achieved a botanical garden. Giving advice to local NGOs and CBOs also indirectly contributes to environmental education. A wide range of workshops and seminars have already taken place in Kenya either targeting staff involved in BIOTA or the stakeholders connected with Kakamega Forest.

Staff training involves direct funding as well as stipends via the KAAD (Catholic Academic Exchange Service) for Master students and PhD candidates, staying in the country or being linked to a German university. Every subproject is keen to contribute to this excellent program of capacity development via supervision of young scientific staff from Eastern Africa. Especially as the BMBF asks for a sound exit strategy after a total funding period of 9 years, this approach is considered as most sustainable. In the end, the scientists involved from Eastern Africa should be able to continue monitoring of biodiversity as well as, together with forest managers, politicians and stakeholders, and the local communities, to use and conserve the forests in a sustainable way.

2.2 Promoting the use of spatial information

What are the specific contributions of subproject E02 for promoting the use of spatial information in biodiversity research and management of East-African rainforests?

Here, for setting-up of the BIOTA-East-GIS a low-cost approach has been chosen, giving preference to a technically simple but efficient realisation instead of making use of complex database technology that requires expert maintainance. Access to the geodata is provided via the internet (www.biota-africa.org, follow links "BIOTA East Africa" and "Data"). For every single dataset meta-information and a quicklook are offered as well as a supported direct download. While the download requires a password and is thus restricted to the BIOTA community, information on available geodata via meta-description and quicklooks is open to the public. The online geodata catalogue is further supported by WebGIS applications. These allow for a detailed judgement on the suitability of geodata before starting downloading data (WALTER and SCHAAB, 2005). Due to poor internet facilities in East Africa, the counterparts are regularly provided with updates via an offline version of the geodata catalogue. As such, the approach followed seems to be especially appropriate when cooperating with institutions in developing countries.

For training of people from East Africa, we offer seminars on geodata handling and processing at the partner institutions (e.g. at NMK in Nairobi). As compared to such seminars we consider supervision of young scientific staff from Eastern Africa during their studies as more sustainable in the long-term. Within the 2nd project phase a PhD candidate supervised from Germany but registered in Kenya is developing a GIS-based biodiversity research support tool. Specific geo-datasets and functionalities for a straight-forward handling also by non-geoprofessionals are to be provided. Besides, two Kenyans are currently taking part in the two-year International Masters Programme in Geomatics at Karlsruhe University of Applied Sciences (Hs KA). Funded by the KAAD, their integration into BIOTA is ensured by student assistance jobs throughout the two years as well as a final thesis within BIOTA-E02.

In order to let the local population better participate in the project outcomes, specific visualisations are planned. These require specific strategies as the local people are not necessarily used to reading maps. We plan to combine video sequences following a certain path or road in the forest or farmland with a map or a satellite image on which the same path is tracked. This is done in order to help to understand the 'view from above'. Testing an English version of the BIOTA-East multimedia presentation (currenly designed and programmed for a museum exhibition, see ZIMMER et al., in print) will give insights in tailoring such media for conservation education. Here, the goal is to increase the local people's ecological sensitivity for a sustainable use of their forests.

But capacity development in the use of spatial information is not restricted to our East-African counterparts or the people living adjacent to the investigated forests. It needs to address also the German partners, as each of them can serve as a multiplier regarding the integration of geospatial reference within their specific field of interest. We cooperate with other BIOTA-East subprojects that are keen in analysing their field data including spatial information (see e.g. BERENS et al., subm.; PETERS et al., subm.). Geodata and field observations could even be combined via spatially explicit GIS- and/or remote sensing-based modelling approaches at the landscape scale.

3 Conclusions

BIOTA-East has promised to ultimately come up with recommendations for a sustainable use and conservation of East-African rainforests at the landscape level. The BMBF asks for a sound exit strategy enabling the counterparts to completely take over biodiversity monitoring in the long term and also forest biodiversity conservation. In order to contribute to the achievement of these goals, for the 3rd and last project phase (2007–2010) E02 has planned to finalize the customized BIOTA-East-GIS interface enabling effective biodiversity management by non-geoprofessionals. Further courses on GIS in general as well as those addressing specific users and needs are envisaged. Again, stipends are hoped for to offer training within the International Geomatics Masters programme at Hs KA. Geodata visualization of BIOTA-East outcomes will cover environmental education, information dissemination, as well as scientific documentation. The setting-up of a geodata-viewing and processing unit in close vicinity to Kakamega Forest is considered as the major exit strategy as it ensures the continuous use of the numerous geodata gathered, processed and produced for use in ground-level management (by FD, KWS) and in research (by NMK, or any visiting scientist working in Kakamega Forest). Supporting the restart of participatory forest management planning for Kakamega Forest and active involvement in its process further ensures that our spatial information as well as the major outcomes of the analysis of long-term forest use and disturbance is applied.

What have we learned so far from this German – East-African partnership in the field of biodiversity research? First of all, there is still great potential for biological research analyses that make use of spatial information. International cooperations struggle in general even more with wishes on the one hand and sensitivities on the other. Allowance must be made for difficulties encountered between people of such divergent cultural backgrounds, but if both parties enter with open minds then the differences can be beneficial for both sides. In Eastern Africa a surprising number of geodatasets are available. But a spatial data infrustructure (SDI) is missing that would ensure straight-forward access to the one true dataset instead of chance-encounters with different modified versions. The possibility for tropical research by researchers from Germany can only be offered in exchange for capacity development. Here, researchers from developed countries should not compete via donations. Instead, assistance in self-help is required; asking for self-commitment should lead to a more sustainable building of capacity. Sadly though, frequent changes of staff at counterpart institutions in Eastern Africa mean that considerable time is consumed by having to make fresh starts for important working relationships. Finally, we argue for a preference for functioning, low-cost approaches instead of seldom-used, hightech innovations, as well as the need for long-term commitments in combination with sound exit strategies.

References

BERENS, D., FARWIG, N., SCHAAB, G. and BÖHNING-GAESE, K., Submitted, Exotic Guava Trees are Potential Centres of Forest Regeneration in Kenyan Farmland.

CBD, 2000, Sustaining Life on Earth. How the Convention on Biological Diversity Promotes Nature and Human Well-being. Published by the Secretariat of the Convention on Biological Diversity. http://www.biodiv.org/doc/publications/cbd-sustain-en.pdf.

FOSTER, D.R., 2006, The Importance of Land Use History to Conservation Biology. In GROOM, M.J., MEFFE, G.K. and CARROLL, C.R. (Eds.), Principles of Conservation Biology. 3rd ed. Sunderland (Massachusetts), pp. 204–207.

FULLER, R.M., GROOM, G.B., MUGISHA, S., IPULET, P., POMEROY, D., KATENDE, A., BAILEY, R. and OGUTU-OHWAYO, R., 1998, The Integration of Field Survey and Remote Sensing for Biodiversity Assessment: A Case Study in the Tropical Forests and Wetlands of Sango Bay, Uganda. Biological Conservation, 86(3), pp. 379–391.

GOTTSCHALK, T.K., 2002, A Remote Sensing and GIS-based Model of Avian Species Habitat and its Potential as a Part of an Environmental Monitoring Programme. Materialien Umweltwissenschaften Vechta, 14.

KÖHLER, J., 2004, Was hat Biodiversitätsforschung mit 'nachhaltiger Nutzung' zu tun? Tier und Museum, 8(3), pp. 82–91.

LUNG, T., 2004, Landbedeckungsänderungen im Gebiet „Kakamega Forest und assoziierte Waldgebiete" (Westkenia) – Multispektrale Klassifikation von Landsat-Satellitenbilddaten und Auswertung mittels Methoden im Raster-GIS. Karlsruher Geowissenschaftliche Schriften, A, 15 (BIOTA East Africa Report No. 3, ed. by G. SCHAAB).

LUNG, T. and SCHAAB. G., 2006, Assessing Fragmentation and Disturbance of West Kenyan Rainforests by Means of Remotely-sensed Imagery Time Series Data and Landscape Metrics. African Journal of Ecology, pp. 491–506.

MITCHELL, N., LUNG, T. and SCHAAB, G., 2006, Tracing Significant Losses and Limited Gains in Forest Cover for the Kakamega-Nandi Complex in Western Kenya across 90 Years by Use of Satellite Imagery, Aerial Photography and Maps. Proceedings of the ISPRS (TC7) Mid-Term Symposium 2006 'Remote Sensing: From Pixels to Processes' (CD), 8-11 May 2006. ITC Enschede (The Netherlands).

MITCHELL, N. and SCHAAB, G., 2006, Assessing Long-term Forest Cover Change in East Africa by Means of a Geographic Information System. Hochschule Karlsruhe –Technik und Wirtschaft, Forschung aktuell 2006, pp. 48–52.

MÜLLER, U., CONRAD, C. and KRAUS, T., 2004, Analyse der raum-zeitlichen Vegetationsdynamik in Ostafrika unter Verwendung von MODIS-Daten. In STROBL, J., BLASCHKE, T. and GRIESEBNER, G. (Eds.), Angewandte Geoinformatik 2004. Beiträge zum 16. AGIT-Symposium Salzburg. Wichmann, Heidelberg, pp. 484–489.

PETERS, M., SCHAAB, G. and KRAEMER, M., Submitted, Impact of Habitat Fragmentation and Degradation on Swarm-raiding Army Ants.

ROY, P.S. and TOMAR, S., 2000, Biodiversity Characterization at Landscape Level using Geospatial Modelling Technique. Biological Conservation, 95(1), pp. 95–109.

SCHAAB, G., KRAUS, T. and STRUNZ, G., 2004, GIS and Remote Sensing Activities as an Integrating Link within the BIOTA-East Africa Project. Proceedings of the International Symposium "Sustainable Use and Conservation of Biological Diversity – A Challenge for Society", Berlin, 1–4 December 2003. Berlin, pp. 161–168.

SCHAAB, G., LUNG, T. and MITCHELL, N., 2005, Land Use/Cover Change Analyses based on Remotely-sensed Imagery and Old Maps as Means to Document Fragmentation and Disturbance for East-African Rainforests over the Last ca. 100 Years. Proceedings of the International Cartographic Conference 2005, 9–16 July 2005, A Coruña (Spain). CD-ROM.

WALTER, M. and SCHAAB, G., 2005, WebGIS zur Unterstützung von Online-Geodatenkatalogen – Realisierung für das BIOTA-Ostafrika-Projekt. Kartographische Nachrichten, 1, pp. 17–25.

ZIMMER, H., DANZEGLOCKE, M. and SCHAAB, G., 2006, Presenting the Project BIOTA East Africa in a Museum Exhibition – A Multimedia Presentation with Special Regard to Remote Sensing and GIS. In: KRIZ, K., CARTWRIGHT, W., PUCHER, A. and KINBERGER, M. (Eds.), Kartographie als Kommunikationsmedium – Cartography as a Communication Medium (= Wiener Schriften zur Geographie und Kartographie, 17). Wien, Institut für Geographie und Regionalforschung der Universität Wien, pp. 36–39.

Chapter 2

Methodologies and Tools

The issues at hand, methodologies to extract information from data and the tools for dissemination have to be developed, adapted and applied based on a thorough analysis of the problems to be solved. What sounds trivial faces many obstacles during implementation. Methodologies successfully applied in one context most often prove to be inadequate when transferred to another. Next to lacking spatial data infrastructures, the hard- and software as well as the skills to make full use of them do have to exist. Methodologies and tools – more often than not – still come with external experts and may cease to be effective once those experts leave. In addition, the new context for the application may require revisiting the problem situation as different perceptions, cultural and political factors have to be taken into consideration for an appropriate solution.

Web-platform: Using the internet and existing information technology components, the SaDIN information platform disseminates spatial information on waste disposal and groundwater extraction, but also connects local experts from communities, government and research as well as international cooperation partners in Morocco and Germany.

Indicator-based sustainability assessment: A GIS-based indicator framework to monitor sustainability links spatial and socio-economic information to improve town planning in Ho Chi Minh City, Cambodia.

Time series: The timely and accurate assessment of land-cover changes is a pre-requisite for regional planning and reporting for international conventions. For this purpose, time series established from MODIS sensor data have been used for updating national databases in Tanzania.

Vegetation Index: Landsat ETM provides information on irrigated crops to monitor groundwater extraction for agricultural production in Syria. While water demand for irrigation is dramatically increasing worldwide, yields could more effectively be improved by selecting suitable environs for specific crops.

Spatial modelling: The CaNaSTA approach demonstrates the potential for niche identification in tropical terrains. However, agricultural production or even subsistence farming is still hampered in some parts of the world recovering from the aftermath of armed conflicts or natural disasters.

GIS and EO: Removing landmines poses a huge challenge to development and requires spatial information systems to improve the effectiveness of demining operations (test cases from Belgium). Reducing the risk posed by natural hazards is the objective of a landslide information system developed for affected regions in Nicaragua.

The evaluation of the presented cases stresses again that spatial data infrastructures are crucial for the effectiveness of methodologies and tools. Information on data has to be spread effectively. More effort has to be spent on establishing SDIs and on capturing baseline data. The challenges are weak administrations, poorly developed relations to local authorities, unproductive bureaucracy, but also rigid traditional research structures and ineffective organisation of projects. The participants of the conference concluded that an interface between experts/scientists and users (e.g. farmers, planners) has to be established at the start of all system development to avoid redundancies and to assure the appropriateness of demand-driven tools. Most of all, however, there is a strong need for benchmarking methodologies and tools. Not every tool offered holds what is promised and the recipients do not command over the expertise and variety of alternatives to judge on their own. Led by associations (such as the African Association of Remote Sensing of the Environment AARSE) and supported by local and international companies a continuous validation process should be established to provide use cases and disseminate best practices.

SaDIN (Sahel-Doukkala Information Network) – Realisation of an Online Geoinformation System in Morocco

Ralf LÖWNER, Abdellatif SOUHEL, Joachim WÄCHTER, Frank KRYSIAK
and the SaDIN working group

Summary

A critical necessity to achieve international standards in public health, sustainable development and land management is to monitor all natural resources, problems and potentialities of a region.

With the contribution of the LIFE – THIRD Countries financial instrument of the European Community, the SaDIN project created a knowledge pool in the centre of the Chouaïb Doukkali University, El Jadida (Morocco), in collaboration with the Data Centre of the GeoForschungsZentrum Potsdam (Germany) and the private enterprise *smartcube ltd.* (Germany), in order to provide a basis for all kinds of environmental investigations. It comprises tools for data acquisition, sample collection and multidisciplinary analysis with a long-term storage database. It addresses fundamental scientific, environmental and anthropological problems of national importance and social relevance. The aim of the project was a regional real-time geoinformation system with worldwide access via Internet and interdisciplinary dynamic content enabling the communication between scientists, technicians and policy makers.

The scientific objectives of this project focused on the evaluation of the potential of aquifers in the region, which supply an arid, yet increasingly densely populated area. These investigations were accompanied by an epidemiological study on the health situation of the local population, which would be seriously affected by groundwater pollution. Another main focus was based on environmental parameters, such as wildlife, vegetation and illegal waste dumps.

The project has been carried out from January 2004 to December 2005 and a huge amount of data has been entered into the implemented databases. It is the interaction of the different working groups and partners, which make a permanent textual and technical enhancement of all components of the system possible.

The experience and knowledge gained in this project will be a source for technology transfer and policy decisions. This is a pilot project, which could serve as an example for similar projects in other regions of northern Africa and beyond.

1 Introduction

Internet is revolutionizing the technical and economic conditions of human knowledge management and became the most important tool for data and information exchange. In Morocco, as in a lot of other African countries, the "Digital Divide" between Africa and

Europe is remarkable and mainly due to the bad conditions of the infrastructure of information but not less to a deficiency in media skills as a result of limited use of information technology and a lack of training possibilities especially for students, not forgetting the traditional methods of teaching that have to be revised. Many highly qualified scientific studies and research projects are realized in Morocco each year, but their results remain dispersed and unexploited.

A first step to bridge the Digital divide was done in a "bottom up" approach: in a locally delimited area, the Sahel-Doukkala region, a "geoinformatic knowledge pool" was created, based on new technologies, especially Internet. The idea was to start up from a manageable sized focus point, with the local people involved in decision making, research and any kind of environmental activities. With this procedure ("bottom up"), the acceptance of the population could be achieved and the later use of the results warranted. In contrast to a national wide approach ("top down"), the results of SaDIN could be spread out little by little to other regions, taking into account the different needs of the local population.

Between 2004 and 2005, the SaDIN (Sahel-Doukkala Scientific Information Network) project was undertaken in the centre of the Chouaïb Doukkali University, El Jadida (Morocco), with the contribution of the LIFE – THIRD Countries financial instrument of the European Community. This project was performed in collaboration between the Science Faculty of the Moroccan University, the Data Centre of the GeoForschungsZentrum Potsdam (Germany) and the private enterprise *smartcube ltd.*, Berlin (Germany), in order to provide a basis for any kind of environmental investigations, especially in the field of provision and protection of groundwater resources. The composition of the consortium is the result of an existing cooperation between the researchers of the involved organizations and will profit from the existing knowledge of the Sahel Doukkala region.

The aim of SaDIN was the creation of an "information center" at the Chouaïb Doukkali University in El Jadida gathering data and information about environment, water resource management, geology, ecology and health. This enables the university to be integrated in the process of land management, offering environmental services and data to local authorities and decision makers. Also, education of the population concerning environmental problems is required. Therefore, the university could become a regional interface providing scientific data and information management systems, and alongside the data and information itself, the training of experts in these areas, and the creation of human resources was one of the major steps towards implementing such structures.

2 Study area

The region of Sahel Doukkala is a coastal plain along the Atlantic Ocean between the cities Casablanca to the North and Safi to the South. It is a basin belonging to the Coastal Meseta of western Morocco with a surface of about 7,700 km^2. The basin is limited in the North and the East by the Oued Oum Erabiaa, in the West and the Northwest by the Atlantic, in the South by the hills of Mouissate and in the Southeast by the Rhamna massif. It is situated in a typical agricultural environment with the semi-arid climate of northern Africa.

The region is projected to become the secondary most important industrial agglomeration after Casablanca. Its industry is characterized by chemical industries (the complex of Jorf

Lasfar and its harbour). Another important activity in the region is the extraction of marine salt from the evaporating onshore swamps (Oulja). Because of these economic activities, the region has a strong attraction for rural populations from near and far.

Considering the Sahel Doukkala region as an industrial, urban and rural environment, with their typical advantages and problems, the investigation area offers outstanding conditions for a pilot project, whose results could be exploited in a broader national or international context.

3 The working groups responding to the environmental problems

With respect to the environmental problems, SaDIN worked upon data from research in different scientific disciplines like geology, hydrogeology, ecology and epidemiology. The investigations focus on three main problems: water resources, waste deposits and desertification (see Fig. 1).

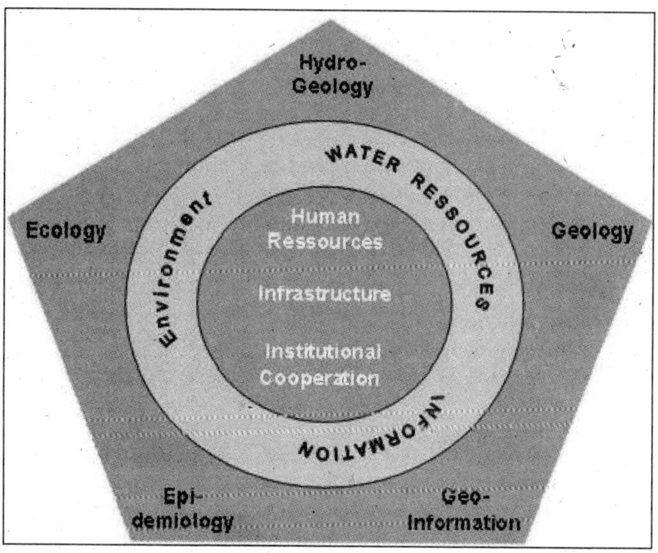

Fig. 1: The working groups and the environmental problems and interests

The water resources of this region are becoming increasingly contaminated due to domestic, industrial and agricultural activities.

First, irrigation degrades the water quality through the process of salinization. The extensive groundwater use for agriculture is responsible for the intense drawdown of the groundwater level. The lower level of fresh groundwater stimulates the intrusion of sea water into littoral zones, which endangers water use and which is responsible for the later degradation of soils and their decreasing agricultural usability. In addition, the extensive use of fertilizer and pesticides also negatively affects the water quality. The extent of these processes is very poorly analysed. But also waste management deficiencies play a major role in water resources. Domestic waste originates primarily from urban agglomerations and is deposited at their periphery. Consequently, the lack of basic infrastructure causes superficial watercourses to transport all types of effluent.

The second main problem is the increasing waste production, due to the rise in socio-economic activities in the Sahel Doukkala region.

The waste is collected and transported in an un-separated manner, throwing together domestic, industrial and hospital waste without treatment. It is deposited in uncontrolled and unprotected dumpsites, which often exceed their capacity with no possibilities for treatment or recycling. Therefore, these places are often called "illegal waste dumps", although being authorised by government. The investigation area contains four large dumpsites. The largest is situated in the suburbs of El Jadida, a side-effect of the rapid urbanisation. Operational since 1983, it accommodates approximately 125 tonnes/day, respectively 0.9 to 1 kg/ habitant/day. These rates are far in excess of the national average (0.6 to 0.7 kg/ habitant/day). The dangerous effluents resulting from the waste materials contaminate the potable water resources often in direct and visible contact with surface water.

The environmental impacts on the atmosphere and the soil water represent potential health risks to the population, especially those living in close contact with the dumpsites.

Desertification is the third main environmental problem investigated in this project. The Sahel Doukkala is one of Morocco's regions, where the effects of interaction between the semi-arid climate and anthropological activities are clearly noticeable. This stress increases desertification and endangers the ecological and socio-economic equilibriums in general. Another important factor, which intensifies desertification is the extension of soil salinization due to intensive water use, particularly where irrigation is carried out. Ultimately, irrigation results in regression of agrarian production and land abandonment.

4 Scientific results

The aim of each working group was two-fold:

- An archive and overview of already existing data was established. The data sources therein are very various, and local authorities are often included in the process.
- New data was collected in the field and in the laboratories of the Chouaïb Doukkala University, augmenting the already existing data pool.

Concerning water resource problems, new scientific results were essentially recorded from detailed geological investigations and the analysis of over 400 wells equally distributed over the research area. The extracted data permitted analyses of the architecture of the aquifer and of ground water levels, of the distribution of chemical elements in the ground water, as well as a soil and fauna inventory. A model including geological, hydrogeological and geographic referenced data was created, which describes the surface and subsurface of the Sahal Doukkala region. An additional benefit of these investigations was the variety of thematic maps for all aspects of geology, for the aquifers and the distribution of salinity and chemical elements.

This geoscientific research evaluated the human impact of water use, polluted by waste disposal sites and illegal waste dumpsites.

With regard to the problems of waste dumps, the health risk for the endangered population was the focus of this project's scientific research. 800 persons (aged between 20 and 40) were subjected to broad epidemiological investigations, concentrating on women and children. For the first time in this region, scientific data of the frequency of diseases and their potential causes have been recorded and a health balance of the population was established. Investigations on potential genetic modifications within the population due to the living circumstances completed this research. Additionally, education measures for the population on environmental problems and first aid programs for the most effected persons were included.

In order to understand the different problems of desertification in the Sahel region, the actual state of natural vegetation and the land use in general was recorded and analysed. Simultaneous investigations on the coastal area focused on the fauna and the identification of endangered species, considering the hydrodynamic data. A result of this work is an archive of faunal and floristic species and a definition of environmental parameters for this region.

These investigations underline the multidisciplinary character of the project: the data of this diverse research will be integrated at a "data center" permitting analysis, comparison and interpretation leading to added value and further information.

5 The geoinformation system

The Digital Divide and environmental problems hamper the development of the Sahel Doukkala region. In the geoscientific section of the Chouaib Doukkali University we tried to address these deficiencies by creating a "geoinformation knowledge pool" based on new technologies, especially Internet. SaDIN enables the storage, analysis, abstraction, synthesis and structured scientific data management in geology, hydrogeology, ecology and epidemiology. Scientific studies are integrated into a geoinformation system, which can be adapted to the needs of the users of this region. This promotes data access, search and controlled manipulation and the scientific discussion.

The main objectives of the installed system reflect the users' needs to exploit information:

- The access to information is guaranteed to all members of the project and to all authorised researchers, technicians and other interested parties in accordance with copyrights, policies and all security aspects;
- The maintenance of a communication tool for the members of project, the researchers and the technicians;
- The access to complex data, derived and processed data and to interpretations based on this data;
- The regular acquisition of multidisciplinary data and its direct dissemination via internet ("real time information system")

The objective was the creation of a scientific data management center for the region of Sahel Doukkala. It will be a "knowledge pool" for the diffusion of information and of an information technology conforming to European standards.

The geoinformation management follows the strategies developed by GFZ Potsdam and *smartcube*, applied worldwide in ICDP projects (International Continental Scientific Drilling Project, www.icdp-online.de).

6 Three-step-realisation

The realisation of the geoinformation part of the project was a three step process. During all steps, an interactive discussion between the computer scientists, geomatic scientists and the geoscientists allowed a continuous adaptation reflecting the users' needs and an ongoing correction of the models. The geoinformation group, which integrates staff members from Morocco and Germany, performed as an interface between the technology and its architecture at one site and the geoscientists at the other site (see Fig. 2).

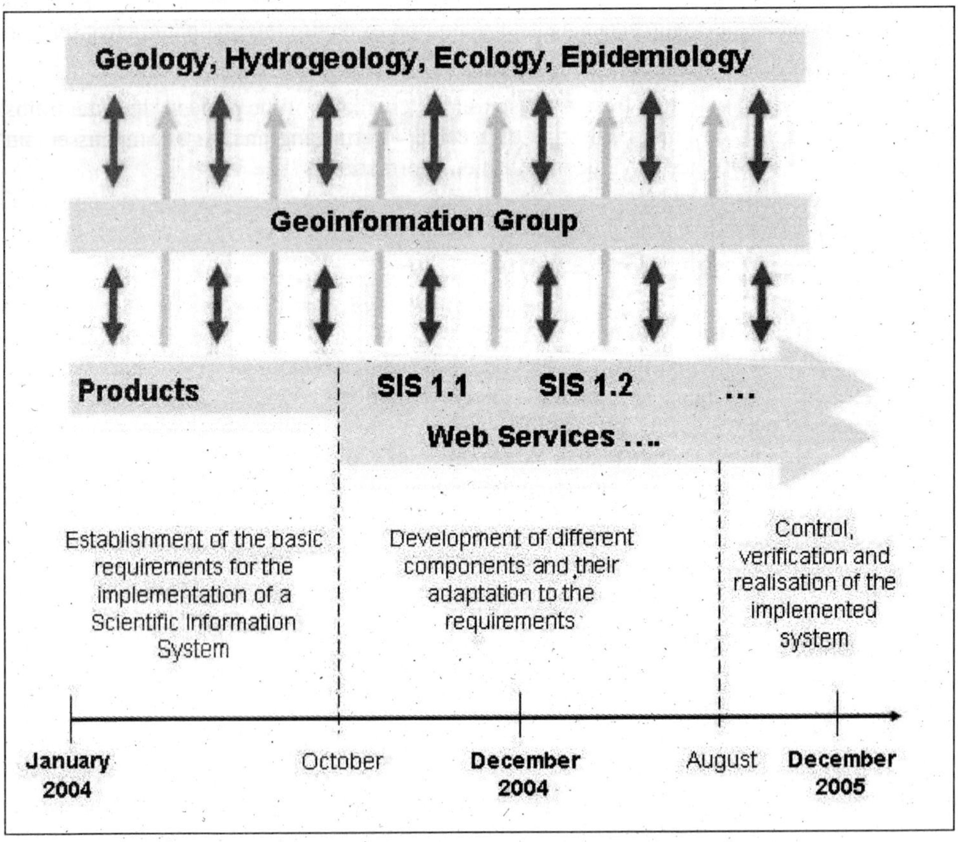

Fig. 2: Workflow and integration of the geoinformation group

In the first step between January and October 2004, multiple analysis of the existing data and of the needs and requirements of the users was undertaken. The resulting user specifications are the foundation for the following investigations. At the same time a web presentation (www.sadin.org) and the corporate lay-out of SaDIN was realised.

To achieve the more significant goals of the project, we also had to implement the basic technical requirements. The local area network was functional by February 2004, and integrated in the Network of the Chouaïb Doukkala University. This work was completed with a highly independent and secure local network.

A data structure model was implemented (see Fig. 3) and used for the adaptation of the existing Drilling Information System (DIS). This software, developed by *smartcube* with assistance of the GFZ Potsdam, was used as the basis for the developed Scientific Information System (SIS). The two main problems that had to be resolved were the geospatial reference of all data and the extremely high multi-disciplinarity, both attributes are not applicable to drilling projects. The developed SIS is a system for data entry, storage, web-publication and visualisation, and had been made functional since October 2004 in the internal Network of SaDIN. It is an open system, which could always be expanding by other research groups, e.g. from social sciences.

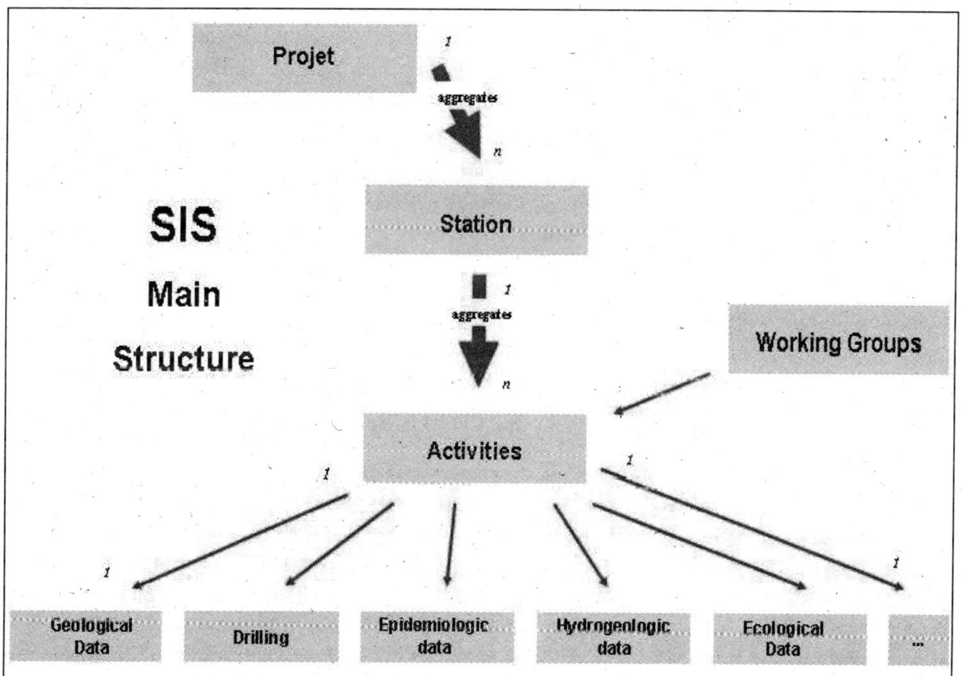

Fig. 3: Superior data structure of the Scientific Information System (SIS) data model

The second step between November 2004 and August 2005, focused on the elaboration of further models and additional services. The main issue was the integration of the SIS in an overall data structure with a data warehouse for long time storage (see Fig. 4) and a clearinghouse for data retrieval and visualisation via internet. Different other services were realised, such as map services and transformation services. Additionally, a content management system separating content from logic and layout, was implemented, so that all project members could edit the content of the web presentation remotely. A database for the concerning literature was implemented. We used the open source product "Winisis" from UNESCO.

Finally, a third step between September and December 2005 comprising control, verification and final adaptations led to the realisation of the implemented system and the diffusion of data and information.

Trainee programs in geoinformation for researchers and students accompanied the whole project as well as the progressive formation of human resources qualified in geoinformatics.

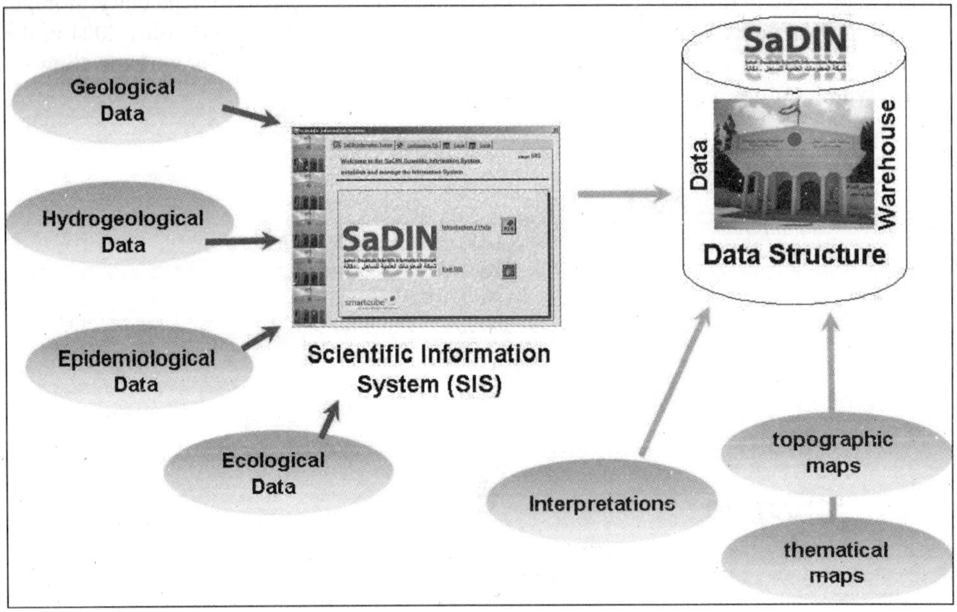

Fig. 4: Integration of the SIS in an overall data structure

7 Three-step-technical overview

The architecture of SaDIN can be divided into the three main components SIS graphical user interface, xSIS web interface and the data warehouse. The whole graphical user interface of SIS is encapsulated in a Microsoft Access Front-End. This Microsoft Access Front End contains all tools to establish, administrate and maintain the SIS. It contains Data-Pumps and Data-Input-Forms for data input facilities through the local Intranet as well as tools for data retrieval facilities. Furthermore it contains special tools to automatically generate and maintain the xSIS Web interface (see Fig. 5).

The Web-Interface-Builder and the Scalable Vector Graphics (SVG) – Builder are used to generate the xSIS components directly out of the SIS. The Web-Interface-Builder is used to generate Microsoft .NET 2.0 Active Server Pages to provide access to the Data Warehouse through the Internet. The appearance and the security mechanism of the Web Interface can be managed in an easy way through this tool.

The SVG-Builder is used to create an interactive map of the region that can be used for visual navigation and data retrieval. Furthermore this tool can be used to generate interactive graphics of the data contained in the Back-End database.

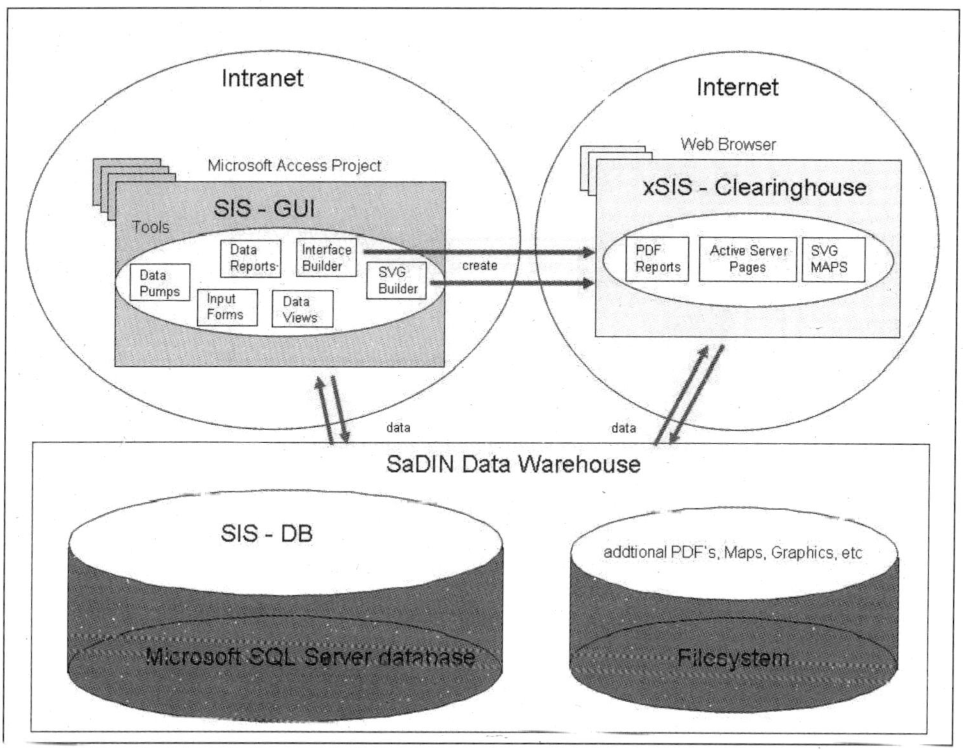

Fig. 5: SIS architecture overview

The xSIS Web interface is build by all data views, data input forms, data reports, graphics and maps that are published by the SIS-Web-Interface-Builder and SVG-Builder (see Fig. 6). Starting from an interactive map of the region most data that is contained in the system can be retrieved by visual navigation. The map contains different layers for each scientific group of SaDIN. Each station and activity is marked on the map and the corresponding data can be retrieved by one mouse-click. Depending on the kind of analysis it is possible to directly open the data view/ report of the analysis or an interactive graphic representation of the data is provided.

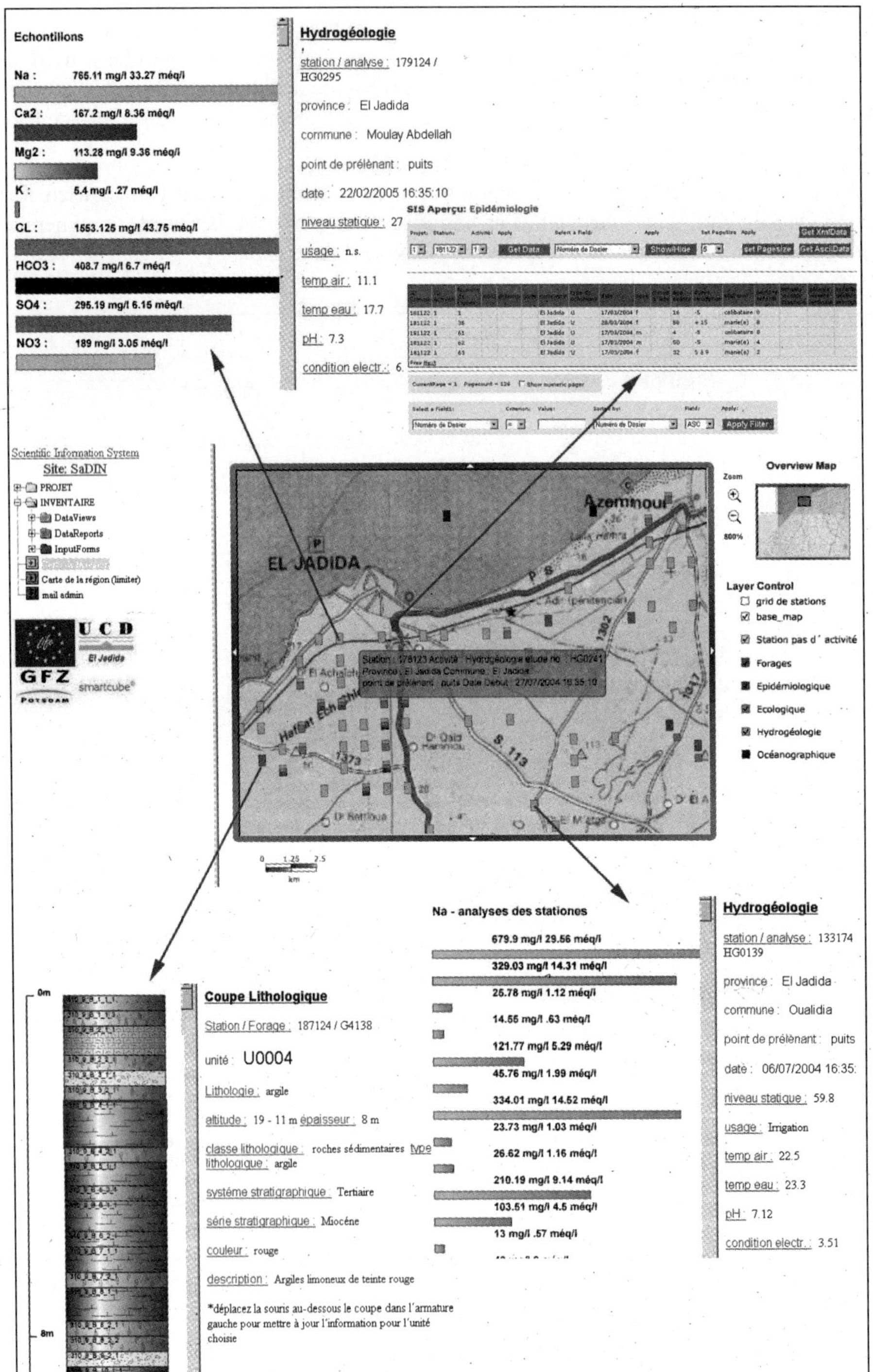

Fig. 6: xSIS Map based Web interface that can be generated by the SIS

The published data views and input forms dynamically request the data contained in the SIS database. All data views and input forms contain facilities to filter the displayed data sets as well as functions to store the data locally in ASCII- or XML-format. To achieve a better performance of the system the data reports are published in static PDF-format.

The Data Warehouse is build by the SIS-Back-End Microsoft SQL Server database and all additional files and graphics that are stored on the SaDIN server. The Microsoft SQL Server database contains all data that is entered through the SIS as well as all additional information and Meta data that is used to administrate and maintain the whole system.

Larger graphics, static PDF-reports and additional files are stored in the file system of the SaDIN server.

8 Future

Starting from these locally restricted results, the implementation of a Spatial Data Infrastructure (SDI) in the Sahel Doukkala region and maybe in Morocco itself could be sparked. SaDIN could work as an internet based communication and information platform (see Fig. 7), which integrates all related disciplines and concerned parties, using a virtual infrastructure. Future work should concentrate on the development of this Spatial Data Infrastructure.

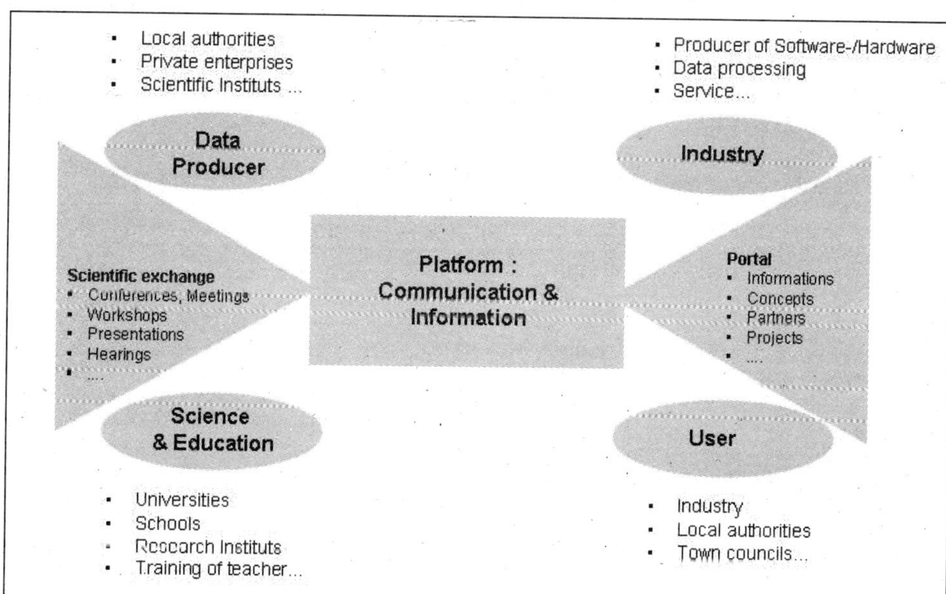

Fig. 7: SaDIN as an internet based communication and information platform

9 Conclusion

SaDIN, a first step to bridge the Digital divide, was carried out in a "bottom up" approach: in a locally delimited area, in the Sahel-Doukkala region, a "geoinformation knowledge pool" was created, based on new technologies, especially Internet.

This project made state-of-the-art technology available in the field of provision and protection of groundwater resources. The investigations resulted in a system, where active and dynamic availability of centralized information is ensured. Different scientific aspects are incorporated in the framework of this project:

- a simulation of different scenarios of future groundwater exploitation and of possible dissemination of salty and polluted fluids into the groundwater;
- an ecological inventory (Fauna and Flora) and possible solutions for the desertification problems for the Sahel region;
- the influence of the environmental situation on the population and on human people in general.

SaDIN enables planning and information management via Internet for future field work, laboratory analysis or other investigations in every research field. It offers the opportunity for cooperation with other organisations (universities, public or private institutions). This project is integrated into the regional land management and used by the local authorities.

The results gained in the SaDIN project could be applied on other regions in North Africa and worldwide.

Acknowledgement

With the contribution of the LIFE financial instrument of the European Community

References

Chouaïb Doukkali University, El Jadida, Morocco:
http://www.ucd.ac.ma
GeoForschungsZentrum Potsdam, Germany:
http://www.gfz-potsdam.de
SaDIN (Sahel Doukkala Sciemtific Information Network):
http://www.sadin.org
smartcube ltd.:
http://www.smartcube.de

GIS-Based Sustainability Assessment of Housing and Settlement Structures Within the Metropolitan Area of Ho Chi Minh City

Harry STORCH

Summary

The paper presents significant initial experiences of a sustainability assessment (SA) research project of housing and settlement policies at the urban planning level in Ho Chi Minh City (HCMC), Vietnam. This research project is financed as part of the new research programme "Megacities of Tomorrow" by the German Federal Ministry of Education and Research (BMBF). The objective is to develop an integrated approach to sustainable development of housing and settlement structures to balance urban growth and redevelopment in HCMC. The initial two-year phase of the project runs from 2005 to 2007. The contribution outlines the methodology of an indicator-based sustainability assessment and strategic environmental assessment (SEA) (SCHMIDT et al., 2005). Urban and regional planning within the metropolitan area of HCMC has been identified as one of the critical policy areas that will need a SA framework to evaluate the effectiveness of planning policies based on quantitative and qualitative baselines for urban indicators. A special focus will be laid on methodological issues of sustainability goals and their spatial representation by urban indicators for the evaluation of housing and settlement strategies. The overall objective of the GIS-based sustainability indicator framework is to promote a better understanding of the environmental and social impacts of planned developments in HCMC.

1 Megacity research within the metropolitan area of HCMC

Asia offers an appropriate setting for the analysis of many of the institutional forces and the urban dynamics that impact the interconnections between humans and their management of environmental resources in the megacities of today (LO and MARCOTULLIO, 2001). Many countries have attempted to limit the growth of their largest cities and have responded to the dynamics of urban explosion with centrally planned institutional and administrative regulations and measures to reduce urban growth. These measures begin with the stimulation of regional and rural development and end with efforts restricting immigration and dispersing new migrants to government-planned resettlement areas as new growth poles. In contrast, solutions seeking to make megacities work for those who are there and the migrants who will inevitably arrive are very uncommon (VAN VLIET, 2002).

Megacities of tomorrow like HCMC offer exceptional opportunities to analyse both the impacts of large-scale environmental resource problems and institutional responses to these impacts, as well as urban planning and management strategies to overcome the limits and failures in the management of environmental resources. The transition of the economic system of Vietnamese cities (BOOTHROYD and PHAM, 2000) has brought about major transformations in the physical and functional urban structures over the last decades. The development of the future megacity of HCMC has two interrelated perspectives: firstly urban growth, the evolving urban forms in the context of urbanisation, and secondly urban redevelopment within the inner urban area. HCMC covers 2000 sq km, divided into 24 districts hosting an official population of more than 6 million. 2 million people have migrated from other provinces and cities, accounting for 30 per cent of the total population. The inner city has an average population density of around 10 000 people per sq km. In HCMC the available average floorspace per capita is 10 sqm. With the expected population-growth and rapidly developing economy, each person's living space will rise to 14 sqm by 2010. HCMC is undergoing a rapid urbanisation such that by 2020 the 17 inner city districts are expected to have a population of approximately 6 million, while the suburban area will have roughly 4 million residents. There is a need for additional housing areas for the resettlement of people living in the canal catchments areas in inner-city districts and for low-income people, civil servants and employees of the armed forces.

This rapid population and economic growth since the policy reform of Doi Moi (*editors:* 'change and newness'; reform and renovation process in the economy of Vietnam, 1986) has put a large and increasing stress on the water resources and environment in HCMC. The demand from industry and households surpassed the current distribution capacities. The water quality in underground sources and river courses is highly degraded due to many sources of pollution. Because the city is situated in an area with plentiful water resources, an efficient water supply should be able to meet the increased demand by urban growth if the water resources are properly developed and managed (VAN DUC and GUPTA, 2000).

In HCMC, sometimes called "motorbike city" with an estimated 0.5 motorcycles per capita, the booming motorcycle traffic is a reflection of its changing living standard. The public transport infrastructure in HCMC can attract only around 10% of travel demand. Statistics show that almost 90% of commuters use private forms of transport. The transportation infrastructure is poor; the density of the road-network is 1 km/km^2 (standard is 5 km/km^2); only 8% of the total urban area is used for the transport infrastructure (standard is 20–25%) and average traffic speed is below 10 km/h. This dominance of motorcycles and the weakness of public transport have resulted in increasing emissions from private urban transport activities. HCMC's infrastructure is overloaded and is unable to meet the needs of people living in highly dense urban areas. It therefore has a negative impact on environment quality and the health of the residents (VAN KHOA, 2001).

The current urban transformation process requires that the urban planning system be based on a sound understanding of the housing and settlement development processes. Housing and settlement development is in essence a complex system with differing patterns of behaviour at various spatial and temporal scales and is centred on understanding the interactions between the built and socio-economic environments and the resulting natural environmental impacts. Urban and regional planning within the metropolitan area of HCMC has been identified as one of the critical policy areas that will need an assessment framework to evaluate effectiveness of planning policies based on quantitative and qualitative baselines for urban indicators. A special focus will be laid on methodological issues of sustainability goals and their spatial representation by urban indicators (RAVETZ, 2000) for the evaluation of housing and settlement strategies.

2 Indicator-based sustainability assessment of settlement and housing developments in HCMC

The research program puts a strong focus on environmental resource management to analyse the settlement dynamics of urban growth and change. The key hypotheses are that different types of environmental and resource problems require particular kinds of management strategies to organise the obvious limits to growth of a pure, informal urbanisation process in metropolitan regions (MITLIN et al., 1996). This sectoral distinction between different management strategies of environmental resource problems is offering the research option to show that urban planning management strategies are more successful in problem solving if they are based on a commonly accepted indicator framework for GIS-based sustainability assessment of housing strategies. Therefore, one important research question is to overcome this traditional distinction between urban planning and information sciences and to analyse the often-ignored relationships between institutions and informed environmental management solutions.

The overall objective of the GIS-based sustainability indicator framework is to promote a better understanding of environmental and social impacts of planned settlement and housing developments in HCMC. In collaboration with university partners from Vietnam (VN), the project has developed a framework for the sustainability assessment of housing strategies, which addresses a broad range of diverse issues. It was with the agreement of the VN partners that the core task of the first stage of the development of an indicator system was designated as the acquisition of key indicator definitions (collection methods, data availability, spatial resolution, currentness, standards and thresholds (legal, political, and administrative). On the basis of this agreed-upon structure, the evaluation of available spatial indicator data according to the following outline has been taking place:

- provision and maintenance of public environmental infrastructure
- protection from environmental pollution (deviation from standards)
- protection, land conservation and sustainable usage of environmental resources.

This bottom-up approach will ensure that the forthcoming development for the indicator-based spatial information system can be built upon available and confirmed data and standards and can be focused on those indicators which exhibit good availability and are relevant to political planning goals and legal standards.

Fig. 1: Sustainable and liveable neighbourhoods: relationships between settlement and housing development and the urban natural environment.

Furthermore, this approach will ensure that relevant stakeholders and administrative and scientific data owners will be able to participate in the project. The indicator framework is derived from the widely accepted and well documented framework in the area of sustainable development. The central concepts of the sustainability indicator framework are housing-related issues and their related environmental and socio-economic indicators. To be applicable, the indicators must have clearly defined reference points beyond the housing-related context.

The following international approaches to urban indicators have been selected and critically assessed for their relevance to housing assessment in HCMC with the help of the regional and local experiences, and the knowledge of counterparts in Vietnam:

- IISD (International Institute for Sustainable Development), because of their published work about indicators for sustainable development with a strong focus on theory, methods, applications (BOSSEL, 1999).
- UN CSD (United Nations Commission for Sustainable Development): Indicators of Sustainable Development – CSD Theme Indicator Framework (UN CSD 2001).
- United Nations Settlements Programme – Guidelines for Monitoring the Habitat Agenda and the Millennium Development Goals (UN-Habitat 2004).
- UNEP (United Nations Environmental Programme) GRID: Encyclopaedia of Urban Environment-Related Indicators (UNEP Grid Arendal 2006).

The selection and prioritisation of these issues (SATTERTHWAITE, 1999), as well as their definition in terms of available or measurable data and their linkage to the overlying indicator framework and their relationships to environmental and socio-economic processes are of importance to a comprehensive sustainability and environmental assessment strategy for housing policies in HCMC.

Table 1: Interrelationship between "liveable housing" and the natural environment described by core sustainability indicators

Liveable housing – Quality of life	
Adequate and affordable housing	**Shelter**
Adequate sanitation Access to clean drinking water Air quality: indoor and ambient Absence of disease vectors Access to health care	**Health**
Threat of natural or man-made disasters	**Safety**
Indicator	*Issue*

Quality of natural environment	
Resource use	Freshwater use Land use Energy use Waste produces
Pollution and degradation	Water quality Air quality
Protection of environment	Conservation (land having formal protection status)
Issue	*Indicator*

Quality of life (Fig. 1), such as access to basic housing needs and public services, and environmental quality are the determining factors in what a sustainable built-up environment and liveable neighbourhood is (SATTERTHWAITE, 1997). However, there is a need for a framework that will be able to support claims about housing needs in relation to the context of society and the physical and natural environment (Tab. 1). The purpose of the research is to investigate the way in which housing needs and sustainability will be able to reinforce each other (GIBBS et al., 2000).

3 Spatialisation of the indicator framework

The complex processes re-shaping settlement patterns in future megacities such as HCMC create urban-environmental systems that cannot be captured by the morphological framework of the urban/rural classifications derived from remote sensing devices. The primary task is therefore to identify key dimensions of regional settlement patterns (SETO and KAUFMANN, 2003), because these dimensions must be represented when moving on to the indicator-centred assessment framework of housing development structures. Based on the evaluation of existing indicators and their refinement with more regionally appropriate measures, the necessary step is to consider how these separated indicators can be explicitly spatially combined in a more multi-dimensional approach to representing sustainability of settlement patterns and housing developments.

3.1 Limits of remote sensing for urban growth research

The urban landscape is dominated by built environments that are physically distinguishable from the surrounding natural environment and therefore are readily identified through the use of remotely sensed image sources (FORESMAN et al., 1997). However, these techniques are often poor when it comes to supporting decision-making on sustainable development, as they are providing detailed descriptions of urban form but are less helpful in understanding the functional characteristics of settlement development processes. Modern remote sensing and GIS techniques have had significant success in quickly monitoring development processes in megacities. Remote sensing and GIS are used in megacity research to provide geo-referenced information on the shape, size and distribution of different land-use classes of the urban environment (PAULEIT and DUHME, 2000).

Even if the technological possibilities of new sensor-technologies are accepted (HEROLD et al., 200[illegible]) there actually remains a remarkable gap between the research-centred results offered by the remote sensing community in the field of urban growth management and the application of these spatial data products by the local planning agencies of urban agglomeration areas. The main reason is that remote-sensing has a strong focus on an ongoing monitoring of urban growth processes, but has its weakness in assessing the environmental impacts and sustainability effects of the underlying urban processes. The framework needed, however, must go beyond the simple identification of an urban environment to integrate the variability of the built environment, which is associated with variability in human use patterns in settlement areas (HARRIS and LONGLEY, 2001).

3.2 Urban typologies as spatial working basis

Different discipline-specific methodological approaches to the 'urban environment' require a commonly accepted spatial working basis, which can ensure that the resulting heterogeneous investigations can be trans-disciplinarily integrated by using an adequate spatially explicit classification. Therefore an 'urban typology' concept was developed and will be used as a practicable method to organise the spatial order of housing developments in HCMC. The housing-related 'urban typology' provides a uniform methodological and spatial framework for the different tasks within the interdisciplinary network of the research project. Housing-related urban development decisions require a rational characterisation of urban structural landscapes according to environmental relevant features. The typology approach ensures that data integration of different sources (remotely sensed, field-based, survey-based and map-based) with their original specific spatial/temporal resolutions and thematic contents can be operationally integrated in the GIS environment of the research-project. The concept of urban-typologies promises to deliver an integrated view of housing-specific problem areas to urban and environmental planners in a form that makes sense to them; because it provides a unit of analysis that is attractive to each of them; because the urban typology creates clusters of residential areas with similar housing structures, service/infrastructure availability, socio-demographic components and environmental pressures; and makes it possible to analyse commonalities and differences in urban development strategies (Tab. 2).

3.3 Indicator-based typology building

There is a need to create urban typologies of the settlement structure of cities such as building/population density, housing types, spread of public services, commute times and other environment-related infrastructure issues. Such an urban typology of housing and settlement structures can take into consideration socio-economic information to determine the liveability and overall sustainability of these individual urban types. The proposed concept represents an interpretative method to integrate the physical aspect of housing developments with the socio-economic and environmental-related information of built-up areas (Fig. 1), based on the concept of urban typologies.

Urban typologies are selective windows on the urban landscape because the typology is based on urban indicators, which direct their focus on particular phenomena and exclude others from the analyses. For that reason, urban typologies and indicators have methodology implications that go well beyond technical discussions of definitions of measurement units. The selection of indicators will directly influence conclusions drawn from the typology-based analysis. The main purpose of urban typologies is to ensure that assessment of planning policies can be clarified and simplified by grouping residential areas and neighbour-hoods with common characteristics and could therefore have similar sustainability problems and environmental impacts. It is important to reflect on the role of urban typologies in bringing clarity to urban planning policy and implementation. This suggests that a limited number of core indicators, based on the key planning policy issues, will be more appropriate than a broader range of detailed descriptions which results in an overly fine-grained geographical classification. This could mean that the application of urban planning policies would become fragmented; thereby negating the principle aim of typologies of defining shared opportunities and providing more consistent and transparent planning approaches.

3.4 Indicator-based multi-layered urban typologies

Because indicators used should reflect the housing-related sustainability issues that the urban typology is seeking to address, a layering of indicators is the most useful approach. It appears to be consensus that a useful urban typology must combine a range of different indicators. The selection of these indicators will directly influence the conclusions that can be drawn from the urban typology approach. The classification of settlement patterns and housing structures should be combined with an analysis of socio-economic, environmental and public service/infrastructure characteristics in order to provide a more accurate picture of current housing problems.

The indicators used to formulate the urban typology are predominantly focused on housing structures and settlement pattern (FLOOD, 1997), with environmental capacity/sensitivity and socio-demographic and economic characteristics also being included. Therefore housing related typologies must be developed on the basis of criteria which reflect these additional socio-economic and environmental issues. The following criteria and indicators have been identified to build the urban typology approach (Tab. 2):

- Urban planning indicators: In urban areas, typologies are intrinsically linked to housing developments. These criteria lead to areas being defined in relation to the characteristics of a city's housing market. This allows for urban planning policies to respond to deficits in housing quality which do not meet the population's needs.
- Spatial environmental planning: Landscape and environmental assessments provide guidance on the ways in which development plans and controls can respond to these issues. Environmental sensitivity and assessment indicators offer the capacity to assist the identification of areas where housing-related development impacts require careful consideration (BOULAND and HUNHAMMAR, 1999).
- Availability of urban infrastructure services: Assessing levels of infrastructure and service is important for the definition of settlement zones. Public service and infrastructure related issues such as access to piped water and sewage capacity, the availability of domestic waste collection services and the lack of these services available to the residents, have an important influence on the ecological footprints and environmental impacts of residential areas (BUTTON, 2002). Transportation-based indicators combine analysis of housing conditions and transport provisions to encourage sustainable modes of transport.
- Socio-demographic characteristics: Population characteristics covering a range of indicators such as social and economic status, and household composition and migration patterns. Economic characteristics can describe levels of dependence on particular sources of employment, income and the relationship between the residential area and the urban economy.

It has been suggested that this urban typology is a more accurate approach that allows for consideration of settlement, socio-economic and environmental characteristics. The ability to develop such an approach is, however, highly dependent on data availability and consistency.

Table 2: Multi-layered Urban Typologies based on Core Urban Sustainability Indicators

Step 1	***Urban Planning Knowledge***
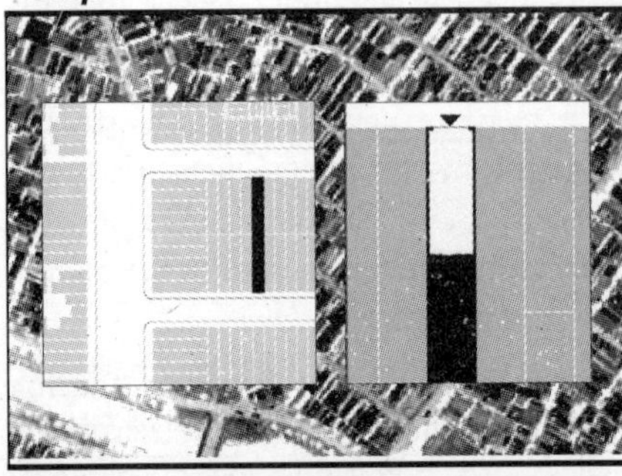	• Urban typologies are linked to housing developments • Housing related typologies must be developed on the basis of criteria which reflect the regional knowledge of urban planners. • Indicators are focused on housing structures and settlement pattern. • Housing structures are detectable with high resolution remote sensing data. • Final definitions need some ground-truthing.
Step 2	***Environmental Planning Knowledge***
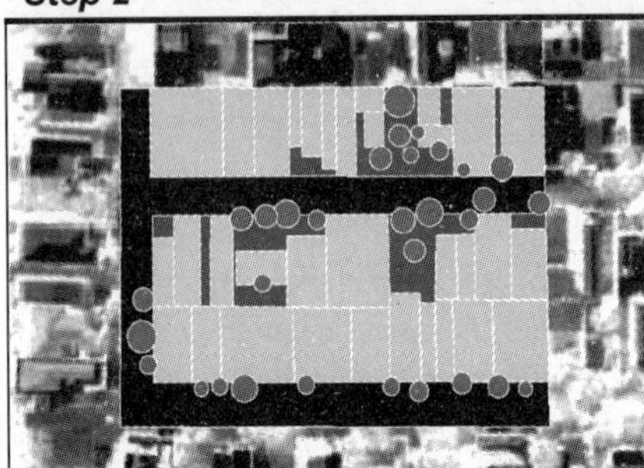	• Linking land-use and vegetation-cover information to housing typologies • Environmentally-based urban administration decisions require characterisation of urban landscapes according to urban land-cover and land-use types. • Integration of available data from different sources is possible (remotely sensed, field-based, and map-based).
Step 3	***Infrastructure/Service-related Knowledge***
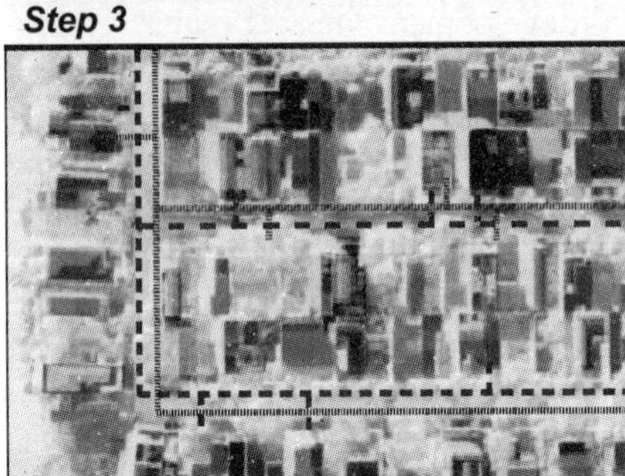	• Availability/Quality of environmental-related infrastructure/services • Assessing levels of infrastructure and service is important for the definition of settlement zones: • Definition of solutions for appropriate urban infrastructure development and sustainable management of urban services to improve the urban environment. • Integration of available data from official sources.
Step 4	***Social Sciences and Human Geography***
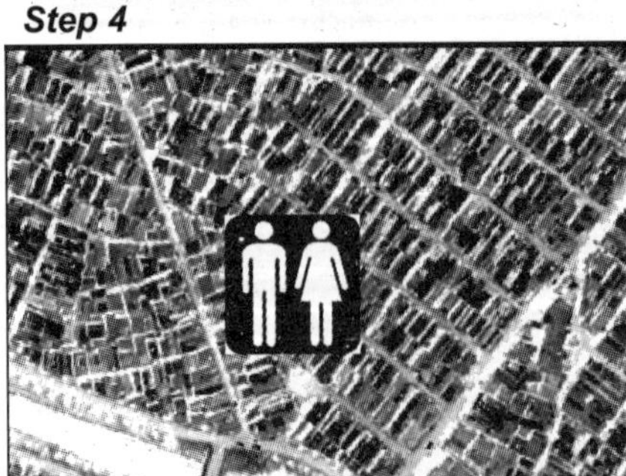	• Socio-demographic and socio-economic information • Socio-demographic indicators such as social and economic status, household composition and migration patterns. • Economic characteristics such as sources of employment, income and the relationship between the residential area and the urban economy.

3.5 Availability of spatial planning data in HCMC

The drawing of conclusions based on a combination of indicators has made clear that some aspects of the urban typology are sometimes difficult to define when data sources available for analyses are lacking in detail – mainly in relation to socio-demographic and public service/infrastructure related issues. These findings reflect the strong influence of settlement and housing strategies on the development of urban typologies. In HCMC, available digital maps are often far from being useful geo-information. Data are stored in different agencies and institutions, in different formats and in different projections. There is obviously a need for the standardisation of GIS data. However, in HCMC the principle challenge in accessing available GIS data is the lack of legal or inter-agency procedures or regulations for data sharing. The availability of documented geographical information in HCMC is currently limited to non-thematic and very general spatial data like topographic maps (base maps), cadastral maps, aerial photographs, satellite images and general thematic maps (like land-use data). However, even these GIS data are sometimes out of date or maintained by and highly distributed among various responsible agencies and data owners. Given that availability limitations and heterogeneity of GIS data in HCMC has exceeded initial expectations, the prototypical implementation of a spatial information system must be limited to case study areas, but the typological approach will permit a sufficient transferability to areas of the city outside of the study areas. Therefore testing the typology in consultation with planning institutions to ensure that it reflects their perception of differences across the urban area will be an important next step. In the case of layered typologies, which combine different indicators, it is important to undertake some ground-truthing.

4 Conclusions and outlook

The strong spatial focus on urban typology facilitates the analyses of different housing-related environmental resource management strategies. Different housing and settlement structures and types have a unique development history, its own built environment representing its specific planning and development history. However, despite these differences, these settlement types have some limiting sustainability factors in common. Regardless of housing structure and the socio-economic development level, all face the challenge of meeting the environmental resource needs of their population on severely limited environmental capacities. All are faced with water resource problems, air pollution and traffic congestion. On a more detailed level, however, it is clear that their planning and management strategies will be clearly distinguishable in their success in solving a number of environmental and resource specific limitations.

The urban typology should be viewed as the missing link between land cover/use centred spatial planning research and urban planning policies (WEILAND, 2001). The urban planning framework is a key mechanism for achieving this in the future, and it must be informed by urban typologies based on core sustainability indicators. This could help the urban planning system to act as an interface, connecting development strategies to produce a more sustainable approach to future developments in the housing sector. This in turn implies that the development of an urban typology requires time and the inputs of different stakeholders if it is to be developed as an administrative urban planning tool that can work over the long term and at different levels. It appears that urban typologies should be sufficiently flexible to allow for the review of change over time.

The outcome of this research project will show whether a GIS-based sustainability assessment of urban developments fulfils only a technical role as a pure planning information system, or whether it might occupy a more central role in terms of sustainability assessment based on indicator-based modelling of the form and function of defined urban typologies. This opens up new prospects for comparing cities in terms of typological-based functional indicators of sustainability, and the integration of the spatial dimension with an understanding of the social, economic, and environmental dimension.

Acknowledgements

The research project 'Sustainable Housing Policies for Megacities of Tomorrow. The Balance of Urban Growth and Redevelopment in Ho Chi Minh City' is financed as part of the new research programme 'Megacities of Tomorrow' by the German Federal Ministry of Education and Research (BMBF). The initial two-year phase of the project runs from 2005 to 2007. The research team is interdisciplinary, and consists of researchers in the areas of urban planning, geography, social sciences and environmental planning (BTU 2006).

References

BOOTHROYD, P. and PHAM, X. N. (Eds.), 2000, Socio-economic Renovation in Viet Nam. The Origin, Evaluation, and Impact of Doi Moi. Institute of Southeast Asian Studies, Singapore.

BOSSEL, H., 1999, Indicators for Sustainable Development: Theory, Method, Applications. A Report to the Balaton Group. Published by IISD International Institute for Sustainable Development. Winnipeg, Manitoba, Canada.

BOULAND P. and HUNHAMMAR, S., 1999, Ecosystem services in urban areas. Ecological Economics, 29(2), pp. 293–301.

BTU (Brandenburgische Technische Universität Cottbus), 2006, Megacities of Tomorrow Ho Chi Minh City/Vietnam. Sustainable Housing Policies for Megacities of Tomorrow. The Balance of Urban Growth and Redevelopment in Ho Chi Minh City. BMBF-Research Programme. http://www.megacity-hcmc.org.

BUTTON, K., 2002, City management and urban environmental indicators. Ecological Economics, 40(2), pp. 217–233.

FLOOD, J., 1997, Urban and Housing Indicators. Urban Studies 34(10), pp. 1635–1665.

FORESMAN, T., PICKETT; S. and ZIPPERER, W, 1997, Methods for spatial and temporal land use and land cover assessment for urban ecosystems and application in the greater Baltimore-Chesapeake region. Urban Ecosystems, (1), pp. 201–216.

GIBBS, D. and JONAS, A., 2000, Governance and regulation in local environmental policy: The utility of a regime approach. Geoforum, 31, pp. 299–313.

HARRIS, R.J. and LONGLEY, P.A., 2001, Data-rich models of the urban environment: RS, GIS and 'lifestyles'. In HALLS, P. (Ed.), Innovations in GIS 8: Spatial Information and the Environment. Taylor and Francis, London, pp. 53–76.

HEROLD, M., HEMPHILL, J., DIETZEL, C. and CLARKE, K.C., 2005, Remote Sensing Derived Mapping to Support Urban Growth Theory. Proceedings URS2005 conference, Phoenix, Arizona, March 2005.

LO, F.C. and MARCOTULLIO, P.J. (Eds.), 2001, Globalization and the Sustainability of Cities in the Asia Pacific Region. New York: United Nations University Press.

MITLIN, D., HARDOY, J. and SATTERTHWAITE, D., 1996, Environmental Problems in Third World Cities. London, Earthscan.

PAULEIT, S. and DUHME, F., 2000, Assessing the environmental performance of land cover types for urban planning. Landscape and Urban Planning, 52, pp. 1–20.

RAVETZ, J., 2000, Integrated Assessment for Sustainability Appraisal in Cities and Regions. Environmental Impact Assessment Review, 20, pp. 31–64.

SATTERTHWAITE, D., 1997, Sustainable cities or cities that contribute towards sustainable development? Urban Studies, 34(10), pp. 1667–1691.

SATTERTHWAITE, D., 1999, The Links between Poverty and the Environment in Urban Areas of Africa, Asia and Latin America, United Nations Development Programme (UNDP) and the European Commission (EC), New York.

SCHMIDT, M., JOÃO, E. and ALBRECHT, E. (Eds.), 2005, Implementing Strategic Environmental Assessment. Environmental Protection in the European Union, 2. Heidelberg: Springer.

SETO, K.C. and KAUFMANN, R.K., 2003, Modelling the drivers of urban land use change in the Pearl River Delta, China: Integrating remote sensing with socioeconomic data, Land Economics, 79(1), pp. 106–121.

UN-HABITAT, 2004, Urban Indicators Guidelines. Monitoring the Habitat Agenda and the Millennium Development Goals. United Nations Human Settlement Programme, Aug. 2004, Nairobi: UN Habitat.

UN CSD (United Nations Commission for Sustainable Development), 2001, Indicators of Sustainable Development – CSD Theme Indicator Framework. http://www.un.org/esa/sustdev/natlinfo/indicators/isdms2001/table_4.htm

UNEP GRID ARENDAL, 2006, Encyclopaedia of Urban Environment-Related Indicators. http://www.ceroi.net/ind/indicat.htm

VAN DUC, L. and GUPTA, A.D., 2000, Water resources planning and management for lower Dong Nai River basin, Vietnam: Application of an integrated water management model. International Journal of Water Resources Development, 16(4), pp. 589–614.

VAN KHOA, L., 2001, Air Quality Management in HO Chi Minh City. Department of Science, Technology and Environment. (DOSTE) of HoChiMinh City – Viet Nam. http://www.unescap.org/esd/environment/kitakyushu/urban_air/city_report/HoChiMinh2.pdf

VAN VLIET, W., 2002, Cities in a globalizing world: from engines of growth to agents of change. Environment and Urbanization, 14(1), pp. 31–40.

WEILAND, U., 2001, Planning Cycle for a Sustainable Development of Urban Regions. Raumforschung und Raumordnung, 59, pp. 392–401.

Thematic Enrichment of LCCS-Based Maps Using MODIS NDVI Data

Federico BREDA, Antonio DI GREGORIO, Fabio MASELLI and Paolo SARFATTI

Summary

The current work aims at providing an operational, automated and easily reproducible methodology to enrich the physiognomic information contained in maps based on the FAO-Land Cover Classification System, LCCS. This is obtained by associating to each map polygon information on vegetation dynamics derived from average NDVI profiles. Imagery of Tanzania from the Terra-MODIS sensor is used for this purpose.

First, the quality of yearly time series of data of the MOD13 product is evaluated in terms of presence of clouds and shadows, residual atmospheric noise, aerosol contamination and other minor effects. Average 23-periods profiles of each polygon of the LCCS based Africover map of Tanzania are extracted, purified, smoothed and converted into 12 periods (monthly) profiles. Their information content on vegetation dynamic is then evaluated. A standardized "a priori" codification system for NDVI profiles and a synthetic codification system for profile clustering are tested.

1 Introduction and objectives

The current work developed and tested a methodology suitable to characterize polygons of FAO Land Cover Classification System maps in terms of average vegetation dynamics (DI GREGORIO and JANSEN, 1996). To pursue this goal, imagery from the Terra-MODIS sensor was used, in particular the product MOD13 that provides NDVI data (*editors*: normalized difference vegetation index). MOD13 products represent a compromise between a good temporal and a fine spatial resolution of respectively 16 days and 250 m, allowing to investigate the average characteristics of vegetation dynamics. The free availability of MODIS data is an additional advantage pushing for their adoption (HUETE et al., 1996; MODIS USER GUIDE, 2003).

NDVI contrasts intense chlorophyll pigment absorption in the red against the high reflectance of leaf mesophyll in the near infrared. This index, which mathematically corresponds to NDVI = [NIR-R] / [NIR+R] (BANNARI et al., 1995), is a widely used index especially when analyzing data taken from satellite platforms. NDVI is indicative of plant photosynthetic activity and has been found to be highly related to the green Leaf Area Index (LAI) and the Fraction of Photosynthetically Active Radiation Absorbed by vegetation (FAPAR) (BARET and GUYOT, 1991; BANNARI et al., 1995). The index can therefore be taken as an optimal indicator of vegetation photosynthetic activity over relatively wide areas, and is intrinsically suited to the current objective.

2 Selection of the study area and data quality evaluation

Four time series of MOD13 data were used, corresponding to the years 2001, 2002, 2003, 2004. Each time series is made of 23 images, each image being a 16 days maximum value composite of daily data with 250 m spatial resolution (BARNES et al., 1998; MODIS USER GUIDE, 2003). The methodology was tested on the MODIS tile h21v09, corresponding to around 90% of the surface of Tanzania (Fig. 1). This area was selected due to its environmental heterogeneity, which makes it representative of a large variety of ecological conditions. Furthermore, the area was suitable for a "worst-case" approach because it shows one to two rainy seasons, during which cloud cover is usually very high producing a particularly problematic situation for the selection of atmospherically undisturbed imagery and the extraction of gap-free NDVI profiles.

Fig. 1: Coverage of the H21V09 tile used as test data set

The methodology was developed and tested by a step-by-step implementation in a PCI Geomatica EASI-PACE procedure which allowed its automatization.

A preliminary image quality assessment was carried out to discard pixels affected by radiometric or atmospheric problems. "Bad" pixels were identified using the NDVI quality layers coming with the MOD13 product, which can be unpacked by the Unpacking Bits procedure code (LDOPE, 2004). Using the NDVI quality layers bad pixels were accounted for due to presence of clouds and shadows, residual atmospheric noise, aerosol contamination and other minor effects (LANDMANN et al., 2005).

Adopting a worst-case approach focus was put on a sub-tile corresponding to South-Eastern Tanzania where higher cloud contamination was expected. The percentages of bad quality pixels were computed as a function of the number of bad 16 days periods within an annual profile for each study year. The same was done for the percentages of pixels as a function of the number of consecutive bad periods. Most pixels contained from 4–5 to 14–15 bad periods, with a certain annual variability whereas the number of the consecutive bad periods generally ranged from 2 to 7–8 (Fig. 2). Thus, only a few pixels were not or marginally affected by the presence of clouds in this sub-zone.

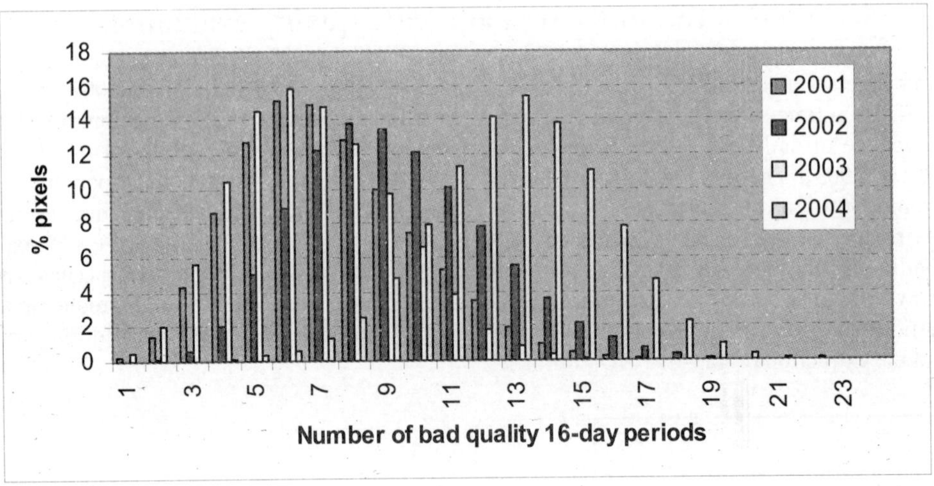

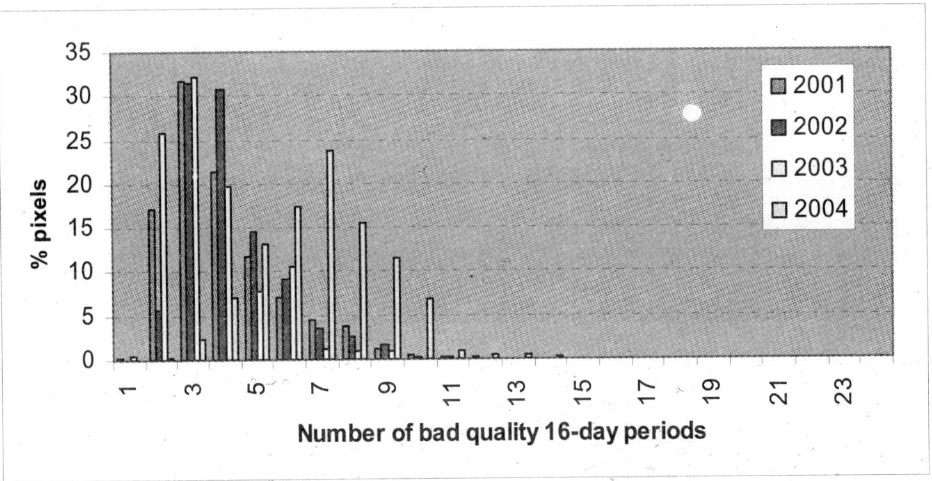

Fig. 2: Percentages of pixels as a function of the number of bad periods within an annual NDVI profile for each study year (above), and percentages of pixels as a function of the number of consecutive bad periods (below)

A method was therefore necessary to reduce the presence of pixels with bad NDVI values and create the most complete 23-period image possible. While making this operation part of the information on annual NDVI, variability was inevitably lost, which was however considered to be of minor importance because the study polygons represented a stable ecological division that was describable only by mean NDVI variability.

A mean NDVI images series was created by averaging the 23 images over the available four years. This was preceded by an evaluation of the number of good values that had to be retained while computing the multi-year average of each pixel. To this aim, mean NDVI image series were computed considering 1, 2, 3 and 4 good NDVI values for each pixel (corresponding to thresholds of bad values equal to 3, 2, 1 and 0). Next, for each image series the percentages of pixels were computed again as a function of the number of bad periods and of the number of consecutive bad periods within the average profile.

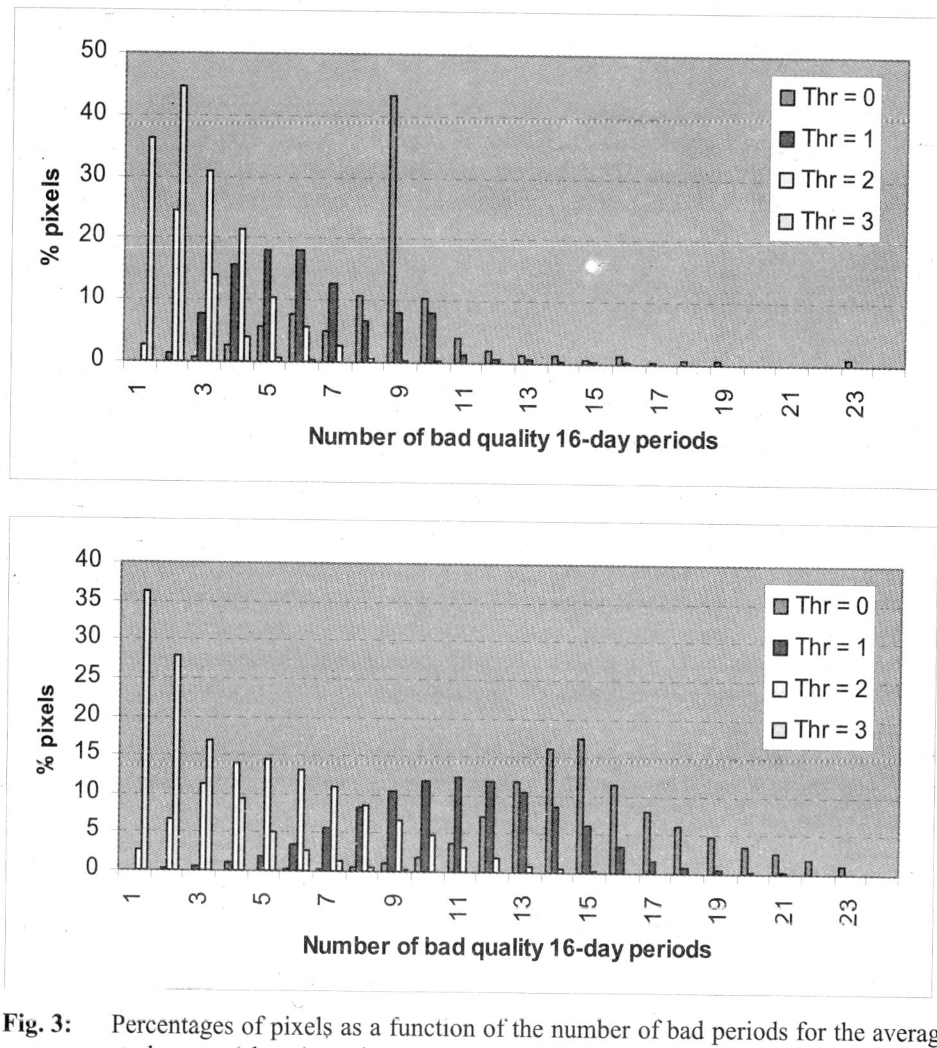

Fig. 3: Percentages of pixels as a function of the number of bad periods for the average study year (above), and percentages of pixels as a function of the number of consecutive bad periods (below). The average study year was computed using different bad values thresholds over the four study years (Thr = 0: all four years were considered; Thr = 3: only one year was considered).

The results of this operation are summarized in figure 3. The application of the lowest quality thresholds (0 bad values) excessively reduced the number of maintained pixels. On the other hand, applying higher thresholds (2,3 bad values) could lead to consider only one year and to produce imagery poorly representative for mean vegetation conditions loosing information on annual NDVI variability.

Consequently, the final choice fell on computing NDVI averages only of pixels with at least two good NDVI values for each 16-day period. Pixels where this condition was not satisfied were discarded. The number of good values averaged for each pixel was retained in an ancillary layer, which also served as binary mask to indicate correctly computed NDVI averages (when this number was greater than one).

Next, a filling of one-period gaps within the 23 average images was performed by replacing the relevant missing values with the local average computed from the previous and subsequent images. In this way, 23 average images with a relatively low number of missing pixels were produced.

3 Extraction of polygon NDVI profiles

Focus was put on evaluating if the multi-temporal profiles derivable from average NDVI images could be used to characterize pure polygons of LCCS maps. Prior to this, some vegetation classes from the map legend of Tanzania developed within the FAO-Africover project were aggregated, obtaining larger classes like herbs, herbaceous crops, tree closed, tree closed needle-leaf and seasonally flooded herbs (FAO-AFRICOVER, 2001).

Exemplary polygons of ecologically relevant classes (herbs, herbaceous crops and trees closed) were selected in different areas of the study tiles. The mean NDVI profiles (23 values) of the polygons containing at least 25 good (i.e. not blank) pixels were extracted. This threshold was chosen considering that it could allow the computation of significant NDVI statistics for each polygon also after eliminating some possible anomalous pixels. Next, the average profiles obtained were visually examined to preliminarily assess their information content on intra-class variability in vegetation development. Exemplary NDVI profiles of polygons of the herbs class showed a high variability: 3–4 NDVI profiles with typical and distinct phenology can be observed in figure 4. A visual examination was performed also on exemplary profiles of herbaceous crops and tree closed classes. The results indicated that several distinct NDVI patterns can be found within each major aggregated class, which confirmed the possibility of using these profiles for intra-class differentiation of vegetation development.

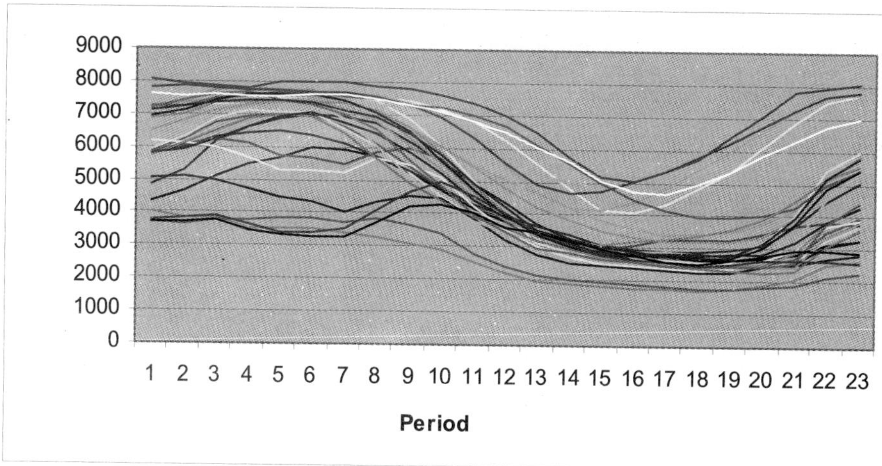

Fig. 4: Exemplary profiles of polygons from class Herbs (the Y scale is in NDVI*10 000)

Since the pure polygon of Africover maps can contain up to 20% of the total surface belonging to other classes (FAO-AFRICOVER, 2001), a method was adopted to remove, or at least reduce, such impurity and isolate the NDVI profiles of the dominant cover type. The method consists of an iterative computation of the NDVI average profile of each polygon, followed by the computation of the Euclidean distance of each polygon pixel from this and the exclusion of the pixel with greatest distance. The process is iterated until a pre-defined number of pixels is removed. The efficiency of the purification was tested on some polygons of herbs, within which homogeneous and pure areas of the majority class were easily identifiable. Thus, root mean square errors (RMSE) were computed between the average NDVI profile of the polygon portions belonging to the majority class and the average NDVI profiles found by applying different levels of purification (i.e. maintaining different fractions of the original pixels) to the corresponding larger (mixed) polygon.

The RMSEs found for four exemplary polygons are summarized in figure 5. These errors are generally decreasing with increased purification, at least up to 20% purification (i.e. maintaining 80% of the original pixels), with the exception of one polygon (herbs 505), which probably contained spectrally irregular surfaces.

The extraction and purification procedure has been eventually applied to all pure polygons of the map obtaining for each of them an average NDVI profile.

The original polygon profiles were still rather noisy. Besides, profiles defined based on 23 periods did not correspond to monthly discrete periods. The profiles were thus smoothed and transformed into 24 periods profiles by using different filtering weights.

Randomly chosen polygon profiles were then extracted for visual analysis. They appeared to be regular and to express information related to the seasonal development of plant photosynthetic activity, and particularly information related to the vegetation's uni- or bi-modal nature and its variation along specific geographic patterns (Fig. 6).

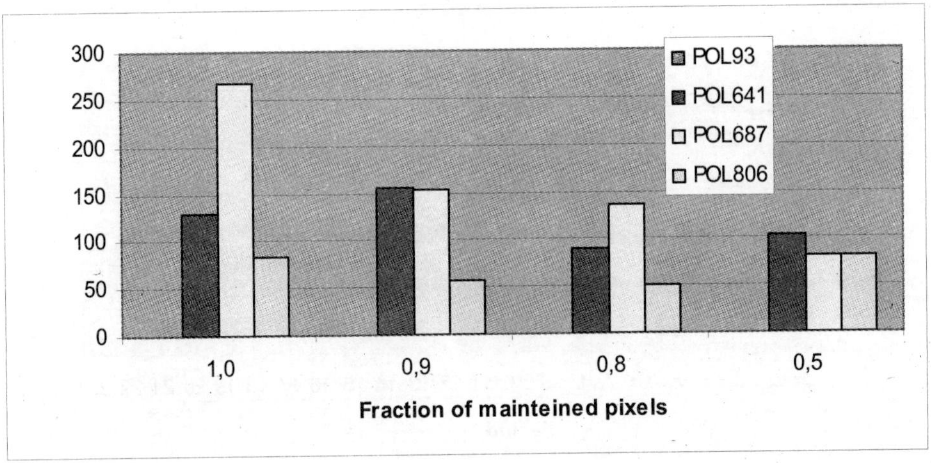

Fig. 5: RMSE (NDVI*10 000) obtained by applying the purification procedure to four exemplary polygons of agricultural areas. The errors were computed with respect to the NDVI profiles of polygon portions belonging to the majority class.

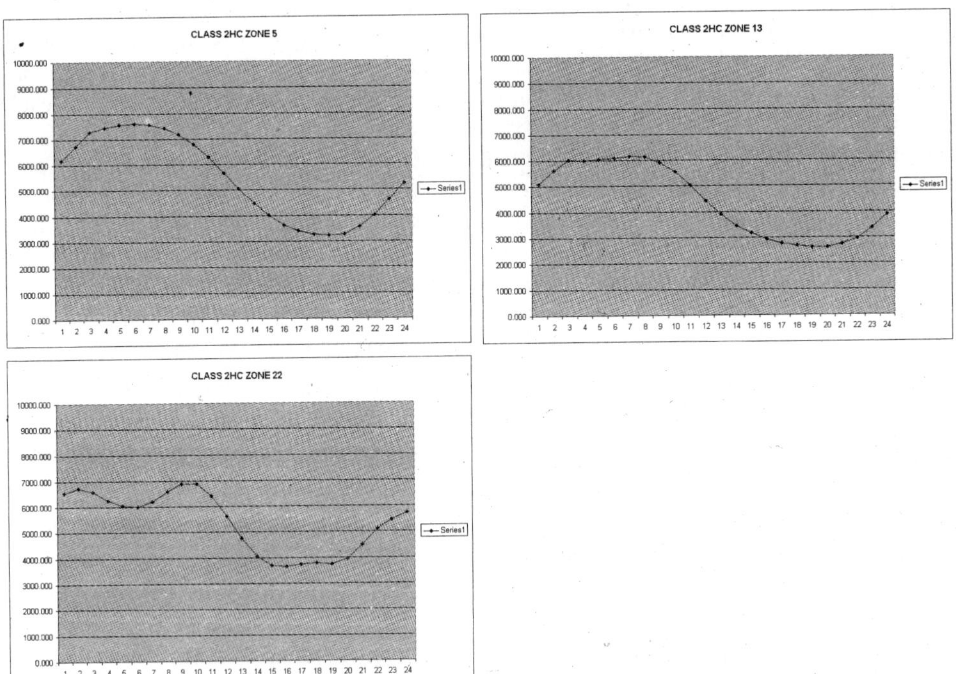

Fig. 6: Average NDVI profiles found within three zones (zone 5 South-East, zone 13 Centre, zone 22 North-West, the Y scale is in NDVI*10 000) placed over a South-East North-West transect

4 Codification of NDVI profiles

An "a priori" codification of polygon NDVI profiles was tested, exploiting the fundamental feature of the NDVI profiles of being completely numeric. This property allowed a complete definition of the code by using a quite limited number of numeric values. The codification has a monthly time step and a NDVI resolution of 0.1. In detail, the NDVI value of each month was multiplied by 10, and the number obtained was approximated to the nearest integer. By this, the values practically always fell between 0 and 9, 12 monthly integer numbers exhaustively described the average annual evolution of a polygon profile. The primary (absolute) NDVI maximum, the possible secondary maximum and the absolute NDVI minimum could be easily indicated within this codification by using bold (for maximum/maxima) and underlined (for minimum) characters (Fig. 7).

This codification presents the advantages of being objective, quantitative and univocal, and it allows simple numerical analysis of the profiles. The complete code of 12 monthly values gives essential information on the average temporal evolution of each polygon NDVI profile.

The codification only suffers from some drawbacks, which are typical for all discretization processes; i.e. the original information of each pair of monthly NDVI values is reduced to a single code of 12 digits. The respective information loss, however, is generally quite limited and it was deemed bearable when compared to less objective codification methods.

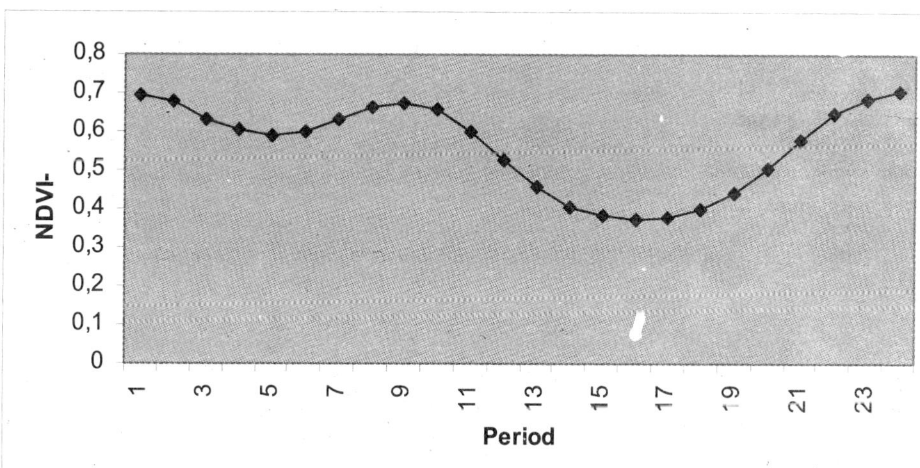

7	6	6	6	**7**	6	4	<u>4</u>	4	5	6	**7**

Fig. 7: Profile of an exemplary polygon and its codification

A further synthesis of the polygon NDVI profiles was also tested, selecting once again an "a priori" codification approach. This codification was based on the following 6 (or 7) classifiers, where all NDVI values are expressed in 0.1 steps (Table 1):

- Number of NDVI maxima (peaks) [generally from 1 to 3];
- NDVI average (0 to 9);
- NDVI maximum (0 to 9);
- NDVI minimum (0 to 9);
- Month of primary maximum (1 to 12);
- Month of minimum (from 1 to 12);
- Month of secondary maximum (if present, from 1 to 12).

Table 1: Synthetic codification of the exemplary profile of Fig. 7

Num of Max	NDVI Average (x10)	NDVI Max (x10)	NDVI Min (x10)	Month of 1st NDVI Max	Month of NDVI Min	Month of 2nd NDVI Max
2	6	7	4	12	8	5

The synthesis of the codification reduced only marginally the information content of the original NDVI profiles. Furthermore, when applied to real polygons from a certain region, it allowed the computation of average polygon profiles useful as reference figures for visually interpreting each eco-climatic situation. Additionally, it facilitated in some cases the clustering of NDVI profiles on the basis of their common codes (Fig. 8).

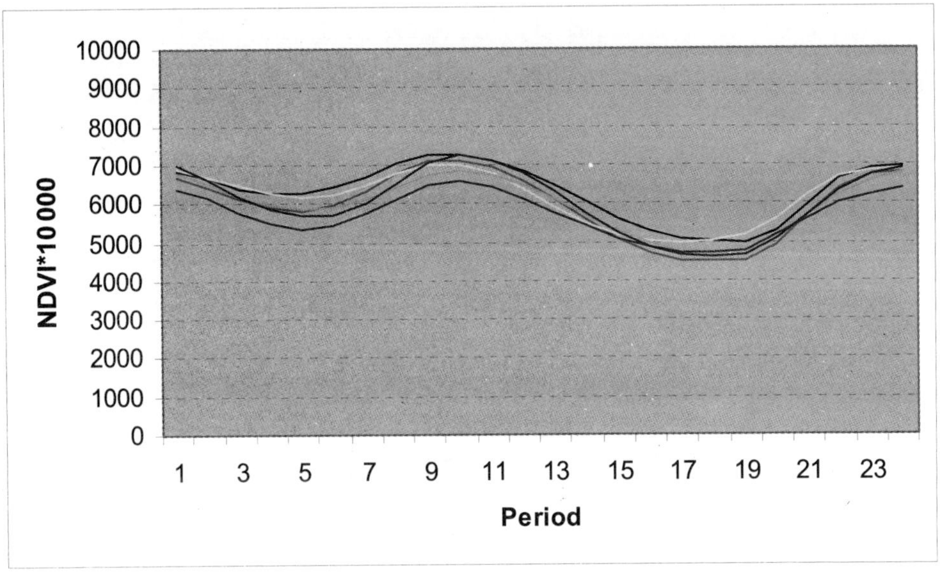

Fig. 8: NDVI profiles of class Very Open Trees with the synthetic code 2-6-7-5-5-9-12

5 Conclusions

The first part of the methodology proved to be easily reproducible and effective in extracting, purifying and smoothing NDVI average profiles for each pure polygon of an input map based on LCCS. The extraction process required several steps of generalization but the profiles maintained a good quality, keeping and expressing their information content in terms of vegetation dynamics.

The two codification methods, which were tested, represented a first attempt to describe concisely the characteristics of the profiles in different eco-climatic situations. The synthetic codification seems to have potential for clustering NDVI profiles according to their common characteristics, but the methods still requires to be fine tuned and further research is needed on this. The codification methods are currently being applied to NDVI profiles of other eco-climatic regions to test whether they are extendable outside of the study area used in this paper.

References

BANNARI, A., MORIN, D., BONN, F. and HUETE, A.R., 1995, A review of vegetation indices. Remote Sensing Reviews, 13, 95–120.

BARET, F. and GUYOT, G., 1991, Potentials and limits of vegetation indices for LAI and APAR assessment. Remote Sensing of Environment, 46, 213–222.

BARNES, W.L., PAGANO, T.S. and SALOMONSON, V.V, 1998, Prelaunch characteristics of the Moderate Resolution Imaging Spectroradiometer (MODIS) on EOS-AM 1. IEEE Trans. Geosci. Remote Sens., 36(4), pp. 1088–1100.

DI GREGORIO, A. and JANSEN, L.J.M., 1996, Land Cover Classification Scheme: A Dichotomous. Modular-Hierarchical Reference System. GCP/RAF/287/ITA Project Doc. No.1/AGLS, Nairobi/Rome.

FAO-AFRICOVER PROJECT, 2001, http://www.africover.org.

HUETE, A., JUSTICE, C. and VAN LEEUWEN, W., 1996, MODIS Vegetation Index (MOD13), Algorithm Theoretical Basis Document (ATBD), version 2.0. NASA EOS Document.

LANDMANN, T., BREDA, F., DI GREGORIO, A., LATHAM, J., SARFATTI, P. and DELLI, G., 2005, Looking Towards A New African Land Cover Dynamics Data Set: The Medium Resolution Data-Base For Africa (MEDA). Proceedings of the AfricaGIS2005 Conference, The Geo-Information Society of South Africa.

LDOPE (2004): Land Distribution Active Archive Centre. http://edcdaac.usgs.gov/tools/modis/.

MODIS USER GUIDE, 2003, http://tbrs.arizona.edu/project/MODIS/UserGuide_doc.php

Geo-Referenced Digital Aerial Balloon Photo Technique – A Low Cost High-Resolution Option for Developing Countries

Jan-Peter MUND and Teng Peng SEANG

Summary

The paper introduces and assesses the highly efficient amateur aerial balloon photo technique which makes aerial photos affordable at a price of 0.10–0.35 US cents per hectare. The idea is to provide an inexpensive but quality efficient solution for recent aerial color photos on a high pixel resolution as orthophotos are not always accessible in Cambodia. Other high resolution images from advanced scanners like Quickbird, Ikonos or others are most likely unaffordable due to insufficient decentralized budget in Cambodia. As a consequence topographical and other spatial data is more or less lacking in planning offices. Enhancement and modifications basically are not easy to integrate into smaller scaled planning resources. This modern technique is based on a simple plastic balloon (r=3m), inflated with hydrogen gas, chemically produced on the site and adapted to the conditions in developing countries especially for areas with rapid land use changes. In Cambodia, it is offered by a private geo-spatial company, the Phnom Penh Geoinformatics Education Centre (PGEC) in collaboration with the Faculty of Land Management and Land Administration at the Royal University of Agriculture.

The rectified, merged and geo-referenced aerial balloon-photos with a surface resolution per pixel of 20 m to 0.4 m, depending on altitude of the platform and camera capacity, can be enlarged up to a scale of 1:800 without any loss in optical and spatial quality. Methodologically the approach uses typical image processing, enhancement and geo-referencing techniques like haze elimination, merging and mosaic procedures, mostly provided by ERDAS IMAGINE software. The aerial photo collection follows standard aerial photo measures like flight line preparation, synchronization of camera and GPS clocks as well as ground truth samples.

The balloon based digital aerial photos from a non calibrated digital camera (Olympus C-70) are already integrated in decentralized regional and urban planning processes in Cambodia as well as geospatial education, data trainings. Further outcomes of this technique are lately renewable detailed land use maps as well as other planning sources for real estate validation and environmental protection strategies i.e. buffer zoning concepts. Besides, they are expected to be useful for other "small area" projects like village based land use planning, technical implementations or (eco-)tourism projects at specific locations and retakes of the same photos within a high-level time resolution of one or two years.

1 Introduction

Digital aerial images provide multi-scale, multi-temporal, and multi-spectral properties thus production of the images becomes a main trend in the field of GIS and photogrammetry (HOFMANN et al., 1984). Geo-referenced or rectified aerial photo image maps are getting important due to its low cost and fast production (JIANN and LIANG, 1999). Accessibility of recent aerial photos with detailed ground resolution of Cambodian environments is sometimes problematic or even impossible for non official or non administrative purposes as well as international collaboration projects. Since 1992 several aerial photos missions have been carried out by contracted international remote sensing companies funded by different international grants and loans. However, most of this spatial data is collected in some Cambodian ministries while a transparent and controlled access or exchange to remote sensing data, even among different ministries is not yet established in Cambodia. In contrast, recent spatial information of fast changing environments is crucial for several administrative, planning, observation and assessment reasons. Interpretation of these high resolution remote sensing data is tempting in the frame of land titling, land use classification, degradation assessments, urban and infrastructure planning as well as monitoring of rapid land cover changes in any kind of environment.

Digital high resolution imagery together with up to date computing power now at nearly everyone's disposal, even in developing countries, has provided a change in scale with more detailed information being sought through the means of digital image analysis (JIANN and LIANG, 1999). For example, it is noticed that highly sophisticated and expensive analog devices are no longer required in photogrammetry if amateur digital camera devices are connected with GPS and GIS image enhancements and geo-rectification techniques (CELIKOYAHAN et al., 2003). Currently several remarkable attempts in balloon born amateur digital photogrammetry have been applied during the last five years (PIVNICKA and KEMPER, 2004). Since two years this technology has been applied in a number of projects on different scale levels in Cambodia (MUND et al., 2005). This paper introduces and assesses the highly efficient amateur aerial balloon photo technique, which makes aerial photos affordable at a price of 0.10–0.35 US cents per hectare, even for low budget projects in developing countries.

1.1 Purpose

Airborne photography and photogrammetry has been a platform for technological improvements and innovations ever since and it was just a question of time to combine balloons, ultra lights and other flying objects such as kites with digital remote observation technology. Major differences between traditional airborne photogrammetry and digital amateur orthophoto gathering technique are the utilization of low budget technology which is easy to transport and handle as well as appropriate under many circumstances with suitable calm weather conditions.

The aim of this project is to incorporate and geo-reference unregistered data from a balloon or ultra light born aerial photography to a coordinate system topographical map. The resolution quality of low-altitude aerial photography far exceeds that of other surface imagery such as Spot or even Ikonos satellite imagery, making it a valuable tool for urban planners, environmental scientists and nature protectionists. Detailed spatial information below 1m pixel resolution such as vegetation associations or housing types is not detectable on inexpensive but larger airborne imagery while high resolution Quickbird images are often not available or cloud covered for Cambodian regions. Detailed construction types and multi-temporal spatial features such as vegetation vitality and small scale clearings are not easily to detect on conventional satellite imagery. Due to geo-referenced aerial photos these features could easily be digitized into separate layers using GIS mapping techniques. However, amateur aerial photos do not generate compass orientation nor any spatial reference information or exact scale. Geo-referencing or ortho-referencing methods needs to be applied, i.e. using referencing tools in ArcMap which allows the user to align geographically unregistered data to data that exists in geographical coordinates.

2 Applied technology

2.1 Types of exercised amateur digital aerial pictures

Two types and various scales of digital aerial photos below 1m pixel resolution are used within the program: geo-referenced and ortho-rectified ones. In most cases mosaic images were produced based on several single pictures using standard remote sensing mosaic production techniques previous to enhancing and referencing the final image. Geo-referenced digital images are transformed to the local co-ordinate system, by warping the digital images to fit either to abase map or other ortho-images. Ortho-rectified picture are processed and shifted to an air triangulation model using the method of surveyed ground control points. Ortho-rectified photos based on digital elevation model techniques are considered to offer higher applicable precision as the accuracy level of simple handheld Garmin GPS 76 can only be improved up to ±2 m applying post processing differential GPS measurements. There are significant benefits in ortho-rectified photography over simple geo-rectified images used here, but there are also significant variations in cost. Ortho-rectified photography has the potential to be more accurate while the photographic image is draped over a digital terrain model based on land survey control. The main benefits are of improved accuracy in horizontal as well as vertical plan shape, particularly where there is high relief across the area. But this accuracy again is strongly dependant upon the accuracy of the DTM used as base model. The most accurate acquirable DTM for Cambodia is the SRTM 50 m grid with provides high spatial resolution with a grid of interpolated levels at 1m intervals. Unfortunately, it has got some significant miscalculations in huge flat lowland areas like the Cambodian floodplain or the Tonle Sap Lake and its tributaries. Only photogrammetrically derived detailed DTM or air triangulation models would provide higher precision and better resolution. This may allow orthophotos to be produced with a plan position accuracy of approx. ±1 m, but was not applied due to limited data processing capacity.

2.2 Hydrogen balloon technique

The utilized balloon is made of light PE material with a maximum size of inflation of about 2.5 m. It is fixed and controlled via 6–8 non flexible offshore fishing lines adjusted at special fixation tongues on the bottom of the balloon. As inflation gas helium or hydrogen can be used. While helium is still very expensive and problematical to acquire in developing countries, hydrogen was chosen for this application in Cambodia. The advantage of hydrogen is its uncomplicated production on the test site based on aluminum cans, water and a special salt which are easy to transport and assembly just beforehand its application (see Fig. 2). The critical danger of hydrogen is its chemical unstable reaction prone to dangerous explosion and inflammation. The chemical reaction formula with the resulting hydrogen gas and Natriumalbumiat is given as follows:

$$AL+NaOH+3H_2O= Na[AL(OH)_4]+1,5H_2O\ .$$

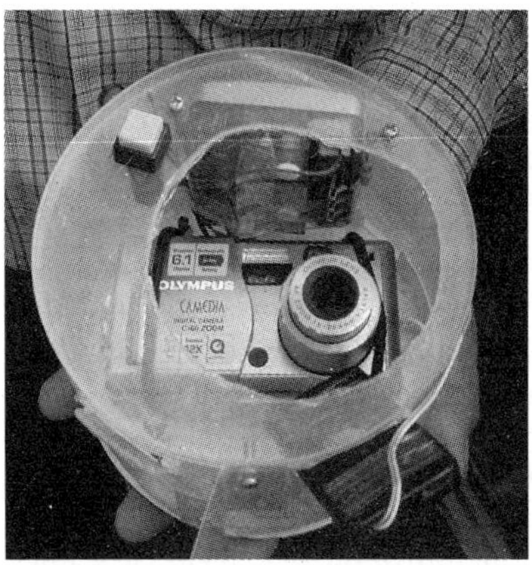

Fig. 1: Camera with GPS receiver

Fig. 2: Balloon filling with Hydrogen Gas

A non calibrated simple 7 Megapixel digital camera (Olympus C70) is fixed in a light plastic holder together with the infrared remote control and a GPS receiver. The camera is activated by a special electronic clock and is set to take pictures every 10 seconds synchronized with a GPS receiver (see Fig. 1). Resulting pictures will have a ground resolution below 1m up to 5–10 cm. Balloon levels of 50 to 400 m above ground with different resulting ground resolution are possible with decreasing positioning accuracy of picture nadir points in higher altitudes. The price of the complete photo acquisition bundle add together around 2500 $ including camera and GPS receiver. Problematic are the balloon sensitivity to wind speed over 3m/sec and explosive inflation gas.

2.3 Field data acquisition

For photo orientation and precise image mosaics ground control points have been measured using post processing differential GPS in comparison to simple hand held GPS measurement. Additionally, a combination of Ground Control Points in combination with image to image rectification technique based on a geo referenced Spot satellite image is considered to provide relative accuracy below average handheld GPS errors. Image orientation and rectification of images and mosaics comparing the two referencing models mentioned above is currently underway. Overlapping aerial photos with 30–40% stereoscopic zones along transects are taken across the target area using a non calibrated digital camera (Olympus C-70) at a high resolution of around 3000×4000 pixels (7.2 megapixels). With a flight altitude of 500–1 200 m this translates to a ground resolution of about 40×40cm or even 10×10 cm per pixel or about 5 to 25 times the resolution of a typical Spot image or aerial photo, which was only available beforehand (see Fig. 3). The resulting image nadir points are corrected with calculated image centers beforehand using the positioning data from GPS receivers and a calculation of real image nadir points and calculated GCPs based on topographical maps of 1:50 000 scale. Plan position accuracy achieved varies according to the variability of ground surface elevation, altitude and the tilt of the camera holder during photography. Plan position is usually found to lie within 2–3 m with raw images but, with full rectification, the expected accuracy ranges in between 0.2 m to 0.5m in horizontal distance to the topographical reference while the vertical accuracy was not yet calculated. Photo files are converted from the Camera TIFF format into World-TIFF format using a six to ten parameter transformation followed by re-sampling of the image using ERDAS software.

2.4 Photo geo-referencing

This rotates and stretch images to match balloon nadir position and corrects for further positioning errors and image distortion. A 1st order transformation will allow the unregistered image to be rotated, shifted, or scaled, but will not deform the image to best match the links. The 2nd Order polynomial transformation allows some distortion of the unregistered image, and should be used if the unregistered image does not quite align with the existing registered map. A 3rd order polynomial transformation is only applied to maximum distortion, or "rubber sheeting", of the unregistered image to match the ground control points. This is necessary for images with larger deformation and where GCP links are poorly aligned after rotation. The majority of balloon images where transformed and re-sampled with 25 to 40 GCPs and a 3rd order transformation method. Transformed and geo-referenced images are then joined into mosaics and ready to be used in further interpretation, mapping and classification processes (see Fig. 6). Mosaic tiles produced in TIFF files, have been found to produce the most versatile images for manipulation within GIS. During mosaic joining cloud areas are masked or extracted and substituted with overlapping areas of neighbouring images.

2.5 Picture scales and mapping techniques

Geo-referenced aerial balloon pictures can be viewed or plotted within a GIS, at scales of better than 1:800 to 1:5000 without any significant loss of image quality. Although a higher level of urban architecture plan shape accuracy can be achieved by mapping from even higher resolutions, the resulting image quality exceeds existing and available conventional aerial photos (see Fig. 4 and 5).

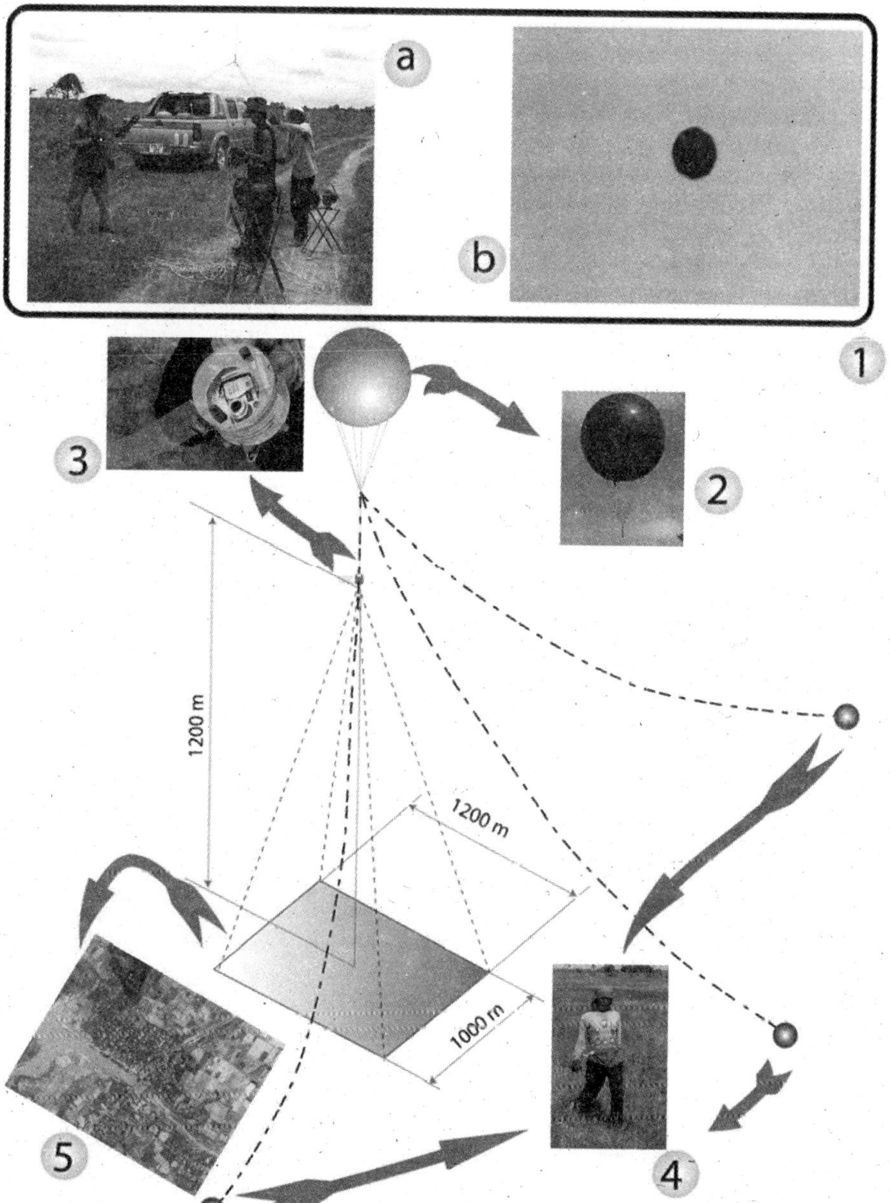

Fig. 3: Model of image acquisition techniquc

Pictures could be even used to count the number of umbrellas of temporary market sellers around market areas. Especially in remote open forest regions the achieved data quality offers new opportunities for inventories and detailed degradation studies. Image data management techniques are applied as follows:

- Photo acquisition and image preparation
- GPS-Data acquisition and post-processing
- Photo mosaics and geo-referencing
- Accuracy assessment
- Final image re-sampling and map production or on screen digitizing

This photo-scale appears to be optimal, for production of geo-rectified images for use in urban and regional planning as well as environmental observations. It produces reliable plan shape data and horizontal resolution for analytical remote sensing data interpretation and on screen digitizing for operational purposes. Even though detailed vegetation or environmental change detection based upon this method is limited, this accuracy is acceptable at all levels of urban planning and change detection of any kind of construction in the urban environment like spontaneous settlements as well as informal construction. Therefore, it is proved to be an appropriate technology for urban planning and construction verification in cities with small to limited budgets.

2.6 Projects using this technique in Cambodia

Digital geo-referenced images of Battambang town (NW Cambodia) were already produced for the district department of urban planning and construction. In 2004 and 2006 aerial balloon surveys were flown on different scales of 40cm and 10cm ground resolution for urban master plan purposes, for regional development as well as change detection in the urban environment with regards to housing and infrastructure measures for informal settlements in the urban areas (see Fig. 6). For a second project, pictures of a scale of 1:10 000–1:15 000 were taken in in remote rural areas with degraded open forest vegetation in NW Kampong Speu province.

Fig. 4: Market Area Battambang 1:1000

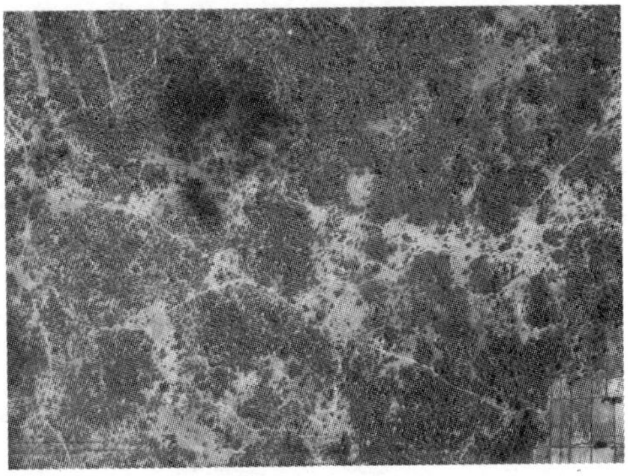

Fig. 5: Degraded open forest area 1:5000

The topic is a seasonal to 5 years change detection observation of a degraded open forest environment for a special observation site of forest re-growth and vegetation recovering (see Fig. 5). Further images were taken for simple area observation and base maps of two national universities in Cambodia (Royal University of Agriculture and Preak Leap National School of Agriculture) in the surrounding of Phnom Penh. But one of the most advantageous applications of this method is the production of base maps for medium scale infrastructure projects like bridge and road construction, hydraulic engineering and irrigation projects in the Cambodian lowlands.

3 Results and discussion

Aerial balloon or ultra-light geo-referenced digital raster aerial photos offer a number of remarkable advantages, when compared with conventional large scale aerial photos and high resolution satellite scanner images. When correctly geo-referenced, these data sets can be used within GIS in combination with vector map data, typically from the topographical maps or field surveying for many planning and observation as well as management applications. Furthermore, digital geo-referenced aerial photographs provide an excellent analytical medium, which can be used conveniently in combination with other types of geo-referenced survey information. In particular, the images provide the opportunity for valuable interpretation of cover changes in vegetation as well as urban land use types by reference to geo-referenced profile data in combination with field surveys.

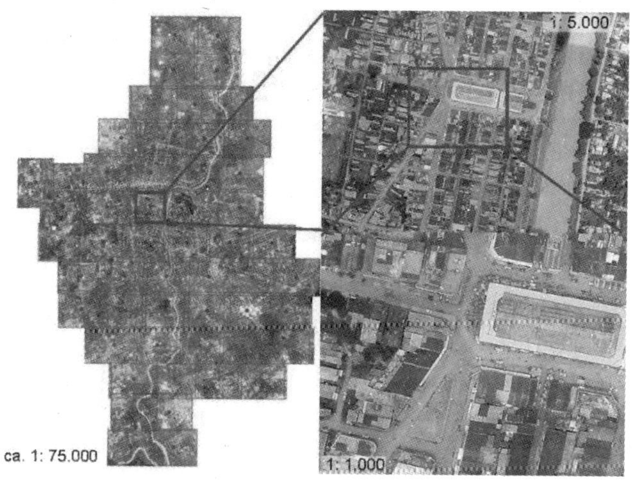

Fig. 6: Image acquisition plan for and detailed Battambang images

Geo-referenced balloon borne aerial digital photos can be an effective replacement for the traditional cost expensive orthophoto images in land cover analysis and urban planning projects and practical for field verification surveys as well. With a non-metric, small, relatively inexpensive digital camera, and inexpensive portable balloon or ultra-light solutions large digital aerial photography is now a suitable technique for developing countries to acquire inexpensive planning information. There is no inevitable requirement for an expensive camera equipment and high-level camera calibration. Once they are geo-referenced, they can be used as a map layers in any GIS and remote sensing application. They can be directly geo-referenced and viewed like maps without extra software for orthophoto re-sampling or other sophisticated image rectification and image enhancement technique. The major advantages of digital aerial photography are that digital images often have much higher resolution than conventional aerial photos. As a result they are efficient both in terms of time and cost. Balloon or ultra-light images offer the following additional detail, relative to conventional orthophoto mapping:

- Urban structure
- Urban infrastructure type an quality
- Detailed land use types in urban and peri-urban area
- Construction plan for medium scale infrastructure projects
- Vegetation classes and vegetation coverage
- Tree density per ha
- Tree types or morphological vegetation structure
- Seasonal and intra seasonal change detection

However, there are several issues to consider before using digital aerial photography like the storage and image representation capacities as well as much higher number of photos for large-scale projects. We acquired more than 84 photos for a quite small region of Battambang town with 2.4 GB of raw data. The technical camera and GPS capacity is even though important for a successful project design as well as GIS and GPS trained staff for ground surveys and data handling. But future digital improvements will further offer enhanced GPS and picture resolution capacity while reducing size and weight of receivers like the new Sony GPS stick. The most challenging issues for digital balloon, ultra-light or kite imagery still remain camera calibration, exact flight management, exact GCP surveys as well as detailed high resolution DTM for ortho-rectification with higher accuracy of processed imagery.

Acknowledgement

The Authors would like to express their thanks to the staff of PGEC Company and students of the Faculty for Land Management and Land Administration of the Royal University of Agriculture in Phnom Penh for their project support. The budget was contributed by the German Development Service (DED) as well as the German Technical Cooperation (GTZ and CIM) and finally the Ministry of Land Management, Urban Planning and Construction (MLMUPC) in Cambodia, as well as the district governor of Battambang have given the administrative permission to carry out several balloon field surveys since 2004.

References

CELIKOYAHAN, M., ALTAN, O., KEMPER, G. and TOZ, G., 2003, Calibrating and using an Olympus camera for balloon-photogrammetry – New perspectives to save cultural heritage. In CIPA 2003 international conference proceedings.

CELIKOYAHAN, M., ALTAN, O., KEMPER, G. and TOZ, G., 2003, Evaluation of a theatre by using low altitude aerial and terrestrial photogrammetry – New perpectives to save cultural hertitage. In CIPA 2003 international conference proceedings.

CHEN, L.C. and LEE, L.H., 1993, Rigorous Generation of Digital Orthophotos from Spot Images. PERS, 59, pp. 655–6611.

CHEN, L.C. and RAU, J.Y., 1993, A Unified Solution for Digital Terrain Model and Orthoimage Generation from SPOT Stereopairs. IEEE Trans. on Geoscience and Remote Sensing, 31(6), pp. 1243–1252.

HAEFNER, S., 2004, The Art of Kite Aerial Photography. Photo Techniques USA Magazine, 7/8 2004.

HOFMANN, O., EBNER, H. and NAVE, P., 1984, DPS – A Digital Photogrammetric System for Producing Digital Elevation Models and Orthophotos by Means of Linear-Array Scanner Imagery. PERS, 50, pp. 1135–1142.

KORPELA, I., 1999, Tree detection from digital aerial photos. An experiment with varying photo and object characteristics. In proceedings of workshop on image analysis and spatial statistics in forestry on November 2, 1999 at KVL, Frederiksberg, Denmark. http://www.dina.kvl.dk/~ml/dina/itcoord/Workshop99/korpela.html.

JIANN. Y. R. and LIANG, C.C., 1999, Digital Ortho-image Generation from Large-Scale Aerial Photography. In GISdevelopment.net, AARS, ACRS 1999, Poster Session 1: http://www.crwr.utexas.edu/gis/gishydro04/Introduction/TermProjects/baldwin.htm; http://www.channelcoast.org/survey_techniques/airborne_remote_sensing_topographic_surveys/?link=digital_aerial_photos.html.

KEMPER, G., 2003, Avionic Systems for application in forestry and agriculture. In 9th conference about Information system for Agriculture and Forestry, proceedings. Prague, CZ.

KEMPER, G., 2006, Spezialapplikationen in der luftgestützten Fernerkundung – preiswerte Systeme, Plattformen und Trägersysteme. In STROBL, J., BLASCHKE, T. and GRIESEBNER, G. (Eds.), Angewandte Geoinformatik 2006. Wichmann, Heidelberg, pp. 296–301.

KEMPER, G., CELIKOYAHAN, M., ALTAN, O. and TOZ, G., 2003, Ballonluftbild- und GIS-basierte photogrammetrische Auswertung kulturhistorischer Objekte in Patara/Türkei. In In STROBL, J., BLASCHKE, T. and GRIESEBNER, G. (Eds.), Angewandte Geographische Informationsverarbeitung IX. Wichmann, Heidelberg, pp. 196–201.

MUND, J.-P., RATTANAK, N. and SEANG, T.P., 2005, Urban and Peri-urban Land Management examples from provincial Cambodian cities. In Proceedings of the FIG SEASC conference in Brunei Daru Salam, November.

PIVNICKA, F. and KEMPER, G., 2004, GIS goes aerial for RS-Data Acquisition. In XXth Congress of the ISPRS. Istanbul.

Groundwater Use in a Dry Region of Syria – Can Satellite Images Be Used to Differentiate Between Irrigated Wheat and Barley?

Wilko SCHWEERS, Huajun TANG, Gerd RÜCKER and Armin RIESER

Summary

In dry regions like the Middle East, irrigated agriculture is consuming a major proportion of the available groundwater. To assess whether groundwater abstractions in a low-rainfall zone are sustainable, water use was recorded jointly with the farmers and the irrigated area was mapped. The resulting ground truth data set was utilized to test possibilities for remote recognition of irrigated areas. Satellite images at the beginning and the end of the wet season were interpreted to distinguish between irrigated barley and wheat. Using an NDVI threshold, most of the irrigated fields were correctly identified on the images and the two crops could be distinguished. However, non-irrigated vegetation with augmented levels of photosynthesis also showed on the images, which reduced the determination coefficient. To some extent, it is possible to single out non-irrigated areas by their shape. More research on the recognition of areas under supplementary irrigation and the distinction of crops with similar signatures is needed.

1 Introduction

In the Middle East, water scarcity is widespread. The major demand is attributed to the agricultural sector, as in Syria, where agriculture uses 85% of the available water resources (FAO-MAAR, 1999). It also contributes about 32% to the GDP, and employs nearly 31% of the workforce, with another 50% of the manufacturing force dependent on agriculture for employment (SALMAN and MUALLA, 2003). The irrigated area has doubled between 1985 and 2003, while the country became self-sufficient in cereals and raised cotton production. More than half of the irrigated area is meanwhile supplied by groundwater. This has resulted in widespread groundwater depletion and deterioration of groundwater quality. However, problems related to groundwater depletion are not limited to the Middle East, but prevail in other parts of Asia, for example in China (JIN et al., 2005) and globally (SHAH et al., 2000).

In one low-rainfall area near the Syrian steppe, groundwater abstraction by farmers was monitored (SCHWEERS et al., 2004a) and evaluated with respect to productivity, profitability and sustainability (SCHWEERS et al., 2004b). In the study area, the farmers in general were responding to water scarcity by a rational use of groundwater resources (SCHWEERS et al., 2005). The groundwater use study revealed differences in the average total groundwater abstraction per ha of wheat and barley of 30% (SCHWEERS et al., 2004a). Since these two crops had the highest overall water use, recognising and differentiating them on satellite images was considered a possibility for refining large-scale water use estimates. As the water use study had produced a reliable set of ground truth data in the form of irrigated areas mapped by GPS, there was an opportunity to differentiate two crops with a similar response to electromagnetic radiation, but different irrigation requirements.

2 Study area

The Khanasser Valley (Fig. 1) is located in Zone 4, approximately 80 km southeast of Aleppo. It stretches 20 km between the Jabboul salt lake in the north and the border of the Syrian steppe in the south. Basalt plateaus of Tertiary age border the valley in the east and west. The annual average long-term rainfall in Khanasser Valley is 209 mm (BRUGGEMAN, 2005). The deep calcareous soils in the valley have a silt-loam to clay-loam texture with moderate infiltration rates (SCHWEERS et al., 2004b). Rearing sheep is the most important source of agricultural income and land-use is dominated by the barley-livestock system. Irrigation is practiced on around 4% of the cropped area. Most of the irrigation wells have been installed in the early 90s. As a result of a "well boom," the groundwater tables had dropped. In 2001, the prohibition of cotton irrigation from groundwater in Zone 4 was enforced to stop groundwater depletion. Since then, the area planted with olives has expanded and cumin was introduced as a cash crop alternative.

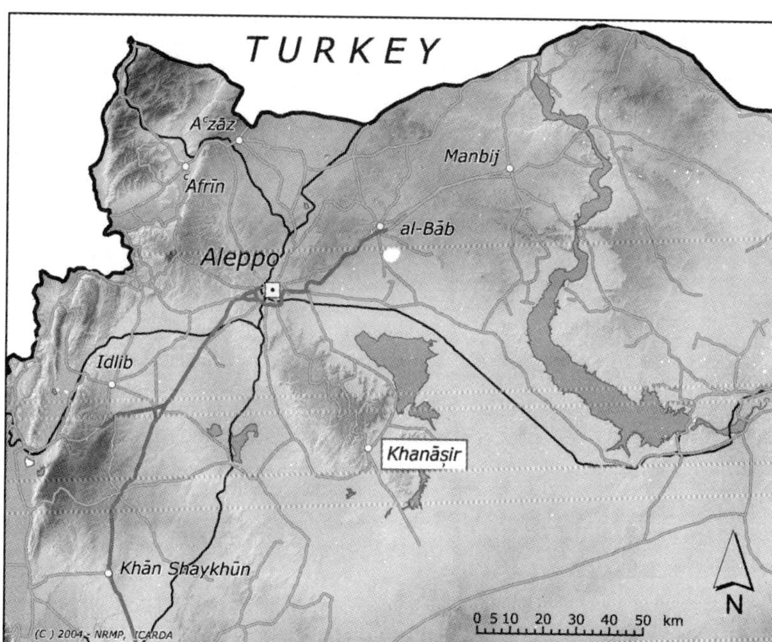

Fig. 1: Overview of Northwest Syria with Khanasser Valley

3 Methodology

Two LANDSAT (ETM+) images were selected. The first image had been taken on November 28, 2002, after barley establishment, the other on May 5, 2003, when wheat was still irrigated. Areas with intense photosynthetic activity, supposed to be irrigated areas, were identified in idrisi32® above a chosen threshold NDVI (barley: 0.18; wheat: 0.24). The NDVI was introduced by ROUSE et al. (1974). It is based on the fact that red light of the visible spectrum gets absorbed in the spongy parenchyma of green vegetation whereas near infrared light is reflected. As LANDSAT sensors measure both spectra, the NDVI can be determined.

$$NDVI = \frac{NIR - RED}{NIR + RED}$$

where:

$NDVI$ = normalized difference vegetation index
NIR = near infrared electromagnetic radiation
RED = visible red electromagnetic radiation

An area of 340 km² was delineated as a subset of the satellite image according to the geographic extension of mapped irrigated fields. The corresponding raster layers of polygons were transformed into vector layers, smoothed with a 3 point low pass filter in idrisi32® and overlaid with the layers of irrigated fields in ArcGIS®. As determination coefficient to measure the accuracy of detection, the ratio between the intersection of the areas above the threshold NDVI and the mapped fields and their respective united area was taken.

$$c = \frac{P \cap I}{P \cup I}$$

where:

c = determination coefficient
P = photosynthetic high activity areas (above threshold NDVI)
I = mapped fields
$\cap$ = intersection operator
$\cup$ = union operator

4 Results and discussion

The timing of satellite images (16-day sun-synchronous orbits) had been appropriately chosen to differentiate between irrigated wheat and barley. Intersections between the NDVI based areas of the two images were negligible. However, irrigated fields were only partially represented by the areas above the NDVI threshold. The incongruence between the supposedly irrigated areas (with high NDVI) and the mapped irrigated areas is attributed to the following reasons:

1. The crops were planted late and the vegetation had not developed beyond the recognition level (some barley fields; see Fig. 2)
2. The farmers had stopped irrigating the crops at an early stage (few wheat fields)
3. The raster and the transformed polygons of above threshold NDVI areas reflected the spatial resolution of the LANDSAT images (30 m; except panchromatic band) which did not correspond exactly to the mapped fields
4. Some irrigated fields, especially in remote areas had been left unmapped (regular shapes)
5. The presence of green vegetation in depressions or in association with water-retaining landscape element, like ancient terraces, bunds, etc. (irregular shapes)

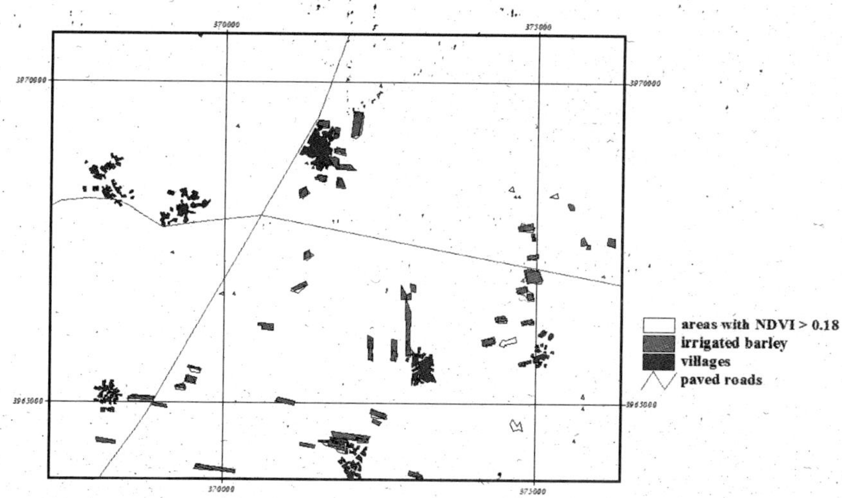

Fig. 2: Areas above a threshold NDVI of 0.18 compared with mapped irrigated barely in the northern part of Khanasser valley

Other irrigated crops were hardly included in the NDVI based areas. Barely is planted in October, about 3 weeks earlier than wheat. Cumin is irrigated by some farmers, but rarely during the last month before harvest. Cotton was traditionally planted after wheat in an annual rotation. If not planted after the same years' wheat crop, it may be irrigated already in April, but since cotton growing was prohibited, in the 2002/03 season the crop had largely disappeared from Khanasser valley. The determination coefficient (Tab. 1) was lower for barely (0.18) than for wheat (0.36). Based on the NDVI threshold maps, irrigated areas were overestimated by 67% (barley) and 60% (wheat). The discrepancy would be somewhat lower taking non-mapped irrigated areas into account. The results are very sensitive to the dimension of the threshold NDVI. In this case, it was chosen by visual comparison with mapped fields. The determination coefficient can also be taken as core element of algorithms for a mathematical optimization of such process. Figure 3 shows the mapped wheat fields superposed on the united area of mapped fields and NDVI.

Table 1: Evaluation of mapped irrigated areas in relation to green vegetation areas indicated by NDVI thresholds

Area (ha)	Wheat	Barley
Mapped	404.5	118.5
NDVI	647.4	197.3
Intersection	279.8	47.1
Union	772.1	268.8
Determination coefficient (c)	**0.36**	**0.18**

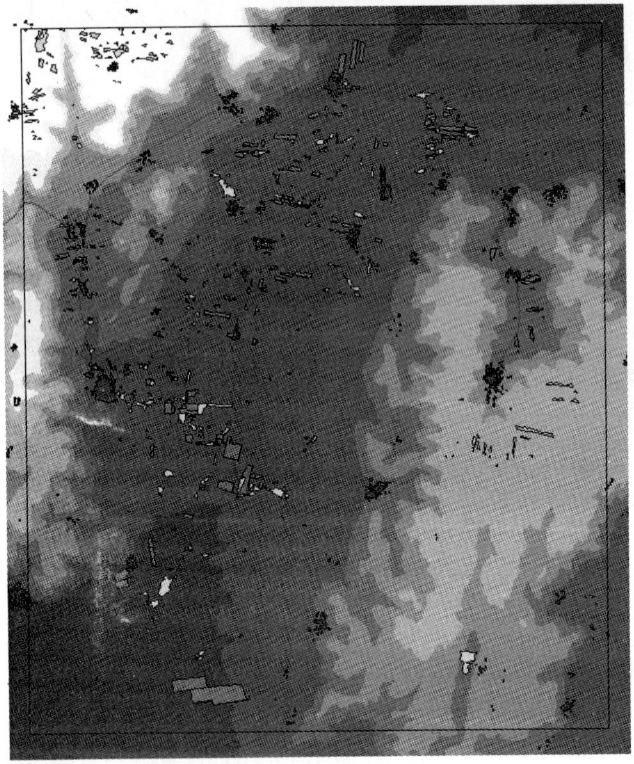

Fig. 3: Wheat fields superposed on the union of mapped fields and areas with NDVI > 0.24 (the frame delineates an area of 17 km × 20 km; vertical difference between elevation lines: 50 m; elevation (h): 100 m asl. < h > 400 m asl.)

NDVI thresholds have been used by other investigators for the determination of irrigated areas, as by SCHWEITZER et al. (2004), who analyzed a combination of images to derive NDVI signatures for characteristic crop rotations in Usbekistan. The NDVI time series were classified into different land uses in a hierarchical decision tree. SAKTHIVADIVEL et al. (1999) made use of seasonal NDVI profiles for differentiating the yield potential of wheat.

They found that in those areas of the Bhakra irrigation scheme, India, where the yield potential was low, the farmers were likely to switch to less water-demanding oilseed crops, thereby releasing more tail end canal water which in that case lead to a rise in water tables and soil salination.

Combining NDVI with information on the shape of polygons might help to increase the accuracy of identifying irrigated areas and separating them from rain-fed areas supplied by surface runoff. This could be done with object-oriented image analysis programs like eCognition®.

4 Conclusions

As a method to separately monitor irrigated barley and wheat, the use of NDVI maps generated from satellite images at different times of the cropping calendar was useful. However, the size of the areas with green vegetation was more than 1.5 times larger than the irrigated areas. To some extent this was due to unrecorded irrigated fields in remote areas. More importantly, rainfall during the winter season had caused the development of green vegetation in runoff areas. Due to their irregular shape, such areas could be separated from the irrigated areas with object-oriented image analysis software. In view of the importance of finding a reliable range of groundwater abstractions in water-scarce regions, an accurate identification of irrigated fields is necessary. The identification of fields with supplemental irrigation is seen as a special challenge.

Acknowledgements

Prof. Peter Zeil and Stefan Kienberger are thanked for their invitation and for their excellent organisation of AGIT 2006 *gi4dev* special session. The crucial file transformation support by Dr. Zhang and Gao Maofang of the MOA key Laboratory of Resource Remote Sensing and Digital Agriculture, Institute of Natural Resources and Regional Planning of CAAS is kindly acknowledged.

References

BRUGGEMAN, A., 2005, Climate of Khanasser Valley. Technical Report. ICARDA, Aleppo, Syria.

FAO-MAAR – Ministry of Agriculture and Agrarian Reform of the Syrian Arab Republic, & Italian Cooperation. Country Profile, 1999, The State of Food and Agriculture in Syria. FAO Project GCP/SYR/006/ITA.

JIN, M., LIANG, X., CAO, Y. and ZHANG, R., 2005, Availability, Status of Development, and Constraints for Sustainable Exploitation of Groundwater in China. IWMI-ITP-NIH International Workshop Creating Synergy between Groundwater Research and Management in South and South East Asia. National Institute of Hydrology, Roorkee, Uttaranchal, India, February 8–9, 2005.

ROUSE, J.W. Jr., HAAS, R., DEERING, D.W., SCHELL, J.A. and HARLAN, J.C., 1974, Monitoring the Vernal Advancement and Retrogradation (Green Wave Effect) of Natural Vegetation. NASA/GSFC Type III Final Report, Greenbelt, MD., 371.

SAKTHIVADIVEL, R., THIRUVENGADACHARI, S., UPALI, W., AMERASINGHE, W.G., BASTIAANSSEN, M. and MOLDEN, D., 1999, Performance evaluation of the Bhakra irrigation system, India, using remote sensing and GIS techniques. Research Report, 28. Colombo, Sri Lanka: International Water Management Institute, 31 p.

SALMAN M. and MUALLA, W., 2003, The utilization of water resources for agriculture in Syria: Analysis of current situation and future challenges. Paper presented at the Erice International Seminars on Planetary Emergencies, August 18–26, 2003, Italy: 11 p.

SHAH, T., MOLDEN, D., SAKTHIVADIVEL, R. and SECKLER, D., 2000, The global ground–water situation: Overview of opportunities and challenges. Colombo, Sri Lanka: International Water Management Institute. http://www.iwmi.cgiar.org/pubs/WWVisn/GrWater.htm#GrRurP.

SCHWEERS, W., BRUGGEMAN, A., RIESER, A. and OWEIS, T., 2004a, Water-use by farmers in response to groundwater scarcity and salinity in a marginal area of Northwest Syria. Journal of Applied Irrigation Science, 39(2), pp. 241–252. http://www.vl-irrigation.org/cms/index.php?id=336.

SCHWEERS, W., OWEIS, T., RIESER, A. and SINGH, M., 2004b, Farmer-participatory research on water use efficiency and irrigation benefit of wheat production in a low-rainfall zone of Syria. Deutscher Tropentag 2004. Berlin, October 5–7, 2004. Conference on International Agricultural Research for Development, Humboldt-Universität, Berlin, 8 p. http:// www.tropentag.de/2004/abstracts/full/393.pdf

SCHWEERS, W., RIESER, A., BRUGGEMAN, A. and MAZID, A., 2005, Groundwater use – assessment and options in a dry region of Syria. IWMI-ITP-NIH International Workshop Creating Synergy between Groundwater Research and Management in South and South East Asia. National Institute of Hydrology, Roorkee, Uttaranchal, India. February 8–9, 2005.

SCHWEITZER, C., RÜCKER, G.R., CONRAD, C., STRUNZ, G. and BENDIX, J., 2004, Knowledge-Based Land-Use Classification Combining Expert Knowledge, GIS, Multi-temporal LANDSAT 7 ETM+ and MODIS Time Series Data in Khorezm, Uzbekistan, 8 pp. In: 1st Göttingen GIS and Remote Sensing Days – Environmental Studies. October 7–8, 2004. Göttingen, Germany.

Spatial Decision Support for Tropical Agriculture – CaNaSTA Probability Modelling

Rachel O'BRIEN

Summary

When identifying suitable niches for specific crops, or identifying the probability of a phenomenon occurring at a specified location, the focus is often on quantifying spatial extent. Whilst this is a fundamental aspect of modelling, it is also important to understand why these locations are selected.

This is specially so for agricultural products for which quality matters, i.e. most higher valued, income generating production alternatives. Furthermore, combinations of spatial and quality modelling will become increasingly important to support denomination of origins for successful product differentiation and product marketing.

Another key issue is deciding how much confidence can be had in model results. This is particularly important when data may be sparse or spatially biased, as is often the case with agricultural trial data.

Spatial decision support can facilitate the decision process by making available relevant data and knowledge. A Spatial Decision Support System has been developed called CaNaSTA (Crop Niche Selection in Tropical Agriculture). The engine of the tool is Bayesian probability modelling. CaNaSTA combines data from trial sites with expert knowledge and environmental data to provide maps, tables and graphs showing where selected species can be expected to thrive, or where a given phenomenon is expected to be present.

CaNaSTA was initially developed as a tool to suggest niche forage species to smallholder farmers in the tropics. Recently, it has also been applied to coffee quality analysis, cowpea performance in tropical hillsides and carbon concentration in soils. The tool is proving useful for analysis of highly specialised crops, where little trial data may be available and variables influencing crop response are often crop-specific. Further research and development aims to provide additional analysis options for specialised crop response research, as well as the application of CaNaSTA to other diverse spatial research problems.

1 Introduction

Research in tropical agriculture faces the challenge of developing crops, technologies and methodologies applicable not only to the unique biophysical environments of the tropics but also to socio-economic niches. Many farmers in tropical countries are smallholder farmers struggling to maintain and improve their livelihoods whilst remaining or becoming self-sufficient. At the same time environmental pressures dictate that technologies are needed which allow farmers to meet these needs in environmentally sustainable ways, through intensification and other appropriate technologies.

A key objective in agricultural research is the application of models for predicting crop performance over large areas. Often little empirical data is available on relatively unknown crops. Trials databases may be biased in a number of ways and are often sparse, making it difficult to extrapolate information of species' response to other locations. The general problem is to identify where these crops would succeed given what trials, experimental data and / or expert knowledge exist.

Lack of information and associated risks form a major constraint to development, so tools that capture data and expert knowledge are extremely valuable.

2 Spatial models for predicting crop performance

Models to predict the response of agricultural crops have long been employed in agricultural research. Agricultural Decision Support Systems (DSS) have been in existence since at least the mid 1970s. There are hundreds of DSS available, covering production decisions relating to crops such as cotton, wheat and pasture (McCown et al., 2002). Increasingly, agricultural DSS are being implemented spatially, and a number of spatial models also exist which can applied to crop performance modelling (O'Brien, 2004). Here, we will concentrate on agro-ecological zones and environmental envelopes, and also consider the role of uncertainty in the modelling process.

2.1 Agro-ecological zones

Agro-ecological zones (AEZ) are environmentally homogenous domains usually derived from climate and soil data (FAO, 1978–1981; Hijmans et al., 2003). The AEZ methodology overlays layers of spatial information to produce zones based on variables such as topography, geology, climate and population. Models are then applied to the resulting zones to analyse land use. The AEZ approach appears to be most suited to well-understood mechanistic models such as crop yield simulation and land suitability.

While this approach is useful for well-studied and broadly adapted crops, it is not immediately applicable to crops where very little is known.

2.2 Environmental envelopes

Species distribution models are generally applied to the natural occurrence of species from an ecology perspective (e.g. Guisan and Zimmerman, 2000). However, the decision of what to plant where (niche modelling) is clearly related. One common algorithm for modelling species distribution is environmental envelopes (e.g., BIOCLIM (Busby, 1991), DOMAIN (Carpenter et al., 1993) and Biomapper (Hirzel et al., 2001)). Environmental envelopes define an envelope in multi-dimensional attribute space within which a species is expected to be found, given known occurrences of the species.

With environmental envelopes, as with many other species distribution models, the response variable tends to be presence / absence, with responses classified as 'present' if they fall within a given percentile for all variables. An additional class of 'marginal' is often added for responses that fall within that percentile for some factors but outside for others, and values within this class may be assigned ranging from 0 (outside percentile for all factors) to 1 (inside percentile for all factors).

In order to specify an environmental envelope, presence points (and absence points, if available) are required. When modelling crop performance, response is less likely to be binary (i.e. presence / absence) and more likely to be ordinal or continuous (e.g. quality, yield or adaptation).

2.3 Uncertainty

Uncertainty arises from ignorance or variability in the decision-making process (FERSON and GINZBURG, 1996), where ignorance describes uncertainty caused by factors that are not considered in a decision, and variability describes the uncertainty that is caused by factors of undefined degree that are known to exist. ROWE (1994) further classifies uncertainty and variability into four main classes, namely, metrical, temporal, structural and translational.

The classifications of Rowe, and Ferson and Ginzburg, provide a useful framework for describing and visualising uncertainty, and examining its impacts in decision-making. When modelling crop response based on trials, it is likely that metrical (spatial and a-spatial) uncertainty, temporal uncertainty and ignorance (lack of data) will be present to a large degree. These uncertainties are difficult to reduce without increased investment in data collection. Therefore, it is reasonable to focus on reducing uncertainty from other sources. Complexity affects both structural and translational uncertainty. Reducing complexity of the model structure reduces associated uncertainty, and because the structure of the model is then also easier to communicate, translational uncertainty is reduced.

3 Tools and methods

The approach used in this research aims to produce a model incorporating both trial data and expert knowledge as inputs, reducing uncertainty where possible and otherwise describing and quantifying it. A Bayesian probability modelling approach is applied to produce maps of responses and associated certainties. This approach has similarities with both agro-ecological zones and environmental envelopes, but differs in that inputs can come from both trial data and expert knowledge, and the model can be specified on a range of responses, rather than presence / absence. In addition, the model can be run dynamically for specific niche crops, and uncertainties are made explicit.

3.1 Bayesian probability modelling

The engine used to develop CaNaSTA is Bayesian probability modelling. Bayesian methods provide a 'formalism for reasoning under conditions of uncertainty, with degrees of belief coded as numerical parameters, which are then combined according to rules of probability theory' (PEARL, 1990). A simple Bayesian model defines prior and conditional probability distributions and combines these to calculate posterior probabilities for each possible outcome. The probability distributions may be derived from data, set by experts or defined from a combination of data and expert opinion.

A 'prior probability' is an initial estimate that may be modified once more information becomes available. If Y is a response variable, then the prior probability of Y is denoted $P(Y)$. 'Joint probability' refers to the probability of two events occurring together, such as a species thriving in a location with certain biophysical conditions. This is denoted by $P(X, Y)$, where X is a predictor variable (e.g., 'rainfall is low') and Y is a response variable (e.g. 'species thrives'). 'Conditional probability' is the probability of a response variable being in a given state, given that a predictor variable is a particular state, and is denoted $P(Y \mid X)$.

Conditional probability can be calculated from prior and joint probability:

$$P(Y \mid X) = \frac{P(Y, X)}{P(X)} \tag{1}$$

It can be shown that posterior probability can be calculated from conditional and prior probabilities:

$$P(Y \mid X^1, X^2, \ldots, X^n) \propto P(Y) \prod_k \left(\frac{P(Y \mid X^k)}{P(Y)} \right) \tag{2}$$

where X^k is the kth predictor variable ($k = 1 .. n$).

Bayesian modelling does not seek to predict exact outcomes, but rather the probabilities of various outcomes, given the effects of the input factors on each outcome. Therefore, if two variables strongly support a given outcome, then the combined effect of these two variables will produce a high probability for that outcome. Conversely, if one variable strongly supports an outcome but another variable does not, then the probability of that outcome occurring will be lessened.

The main assumption in the above approach is that of conditional independence (CI). If two variables are conditionally independent, then knowing the state of one variable has no bearing on the probability distribution of another variable once the state of the response variable is known. The requirement of CI is often not met when dealing with environmental data (ASPINALL, 1992).

The probability of datasets being conditionally dependent can be lessened by using fewer datasets. Moreover, functional independence may be more critical than conditional independence (CORNER et al., 2002). Even if two datasets are statistically dependent (for example, because they are derived from a common source), or if they have different meanings in the model then it may still be valid to include them both.

3.2 CaNaSTA implementation

The CaNaSTA algorithm (O'BRIEN, 2004) creates conditional probability tables of all predictor variable categories against response variable categories. Probability distributions are calculated for a response variable under specific sets of environmental conditions. A response variable will typically be growth or product characteristics of a species, such as quality, yield or adaptation. However, CaNaSTA can also be applied to other modelling problems, such as presence of soil characteristics or other spatial phenomena. The primary model output is a discrete probability distribution at each location and CaNaSTA is particularly suited to ordinal response variable categories. Predictor variables can be any variables for which data is available for the trial sites, and which is considered to be relevant to the model. The main assumption of Bayesian probability modelling is that of conditional independence, i.e. predictor variables should independent when conditioned on the response variable. CaNaSTA includes descriptive statistics to assess the likelihood of conditional independence being violated for the predictors selected.

The CaNaSTA algorithm creates conditional probability tables of all predictor variable categories against response variable categories. If there are a large number of occurrences in the trial data with a particular combination of predictors and responses, then certainty associated with the probability distribution is deemed to be high. Where few or no trials exist, the associated certainty will be low. Where conditional probabilities cannot be determined from trial data, expert knowledge can be used to estimate conditional probabilities, thereby increasing the certainty associated with a particular conditional probability value.

Defining the model in terms of probabilities gives some measure of certainty. However, it would also be useful to know something about the certainty of the probability distribution itself. In the case where probability distributions are derived from frequency counts in a trials database, then a certainty related to the probability distribution can be derived directly from the number of trials. When probability distributions are elicited using expert knowledge, these measures can also be applied, depending on how much certainty the expert associates with their judgement.

From Equations 1 and 2, in order to specify posterior probability distributions $P(Y \mid X^1, X^2, \ldots, X^l)$ joint probability distributions $P(Y, X^k)$ ($k = 1$ to l) and prior probability distributions $P(Y)$ and $P(X)$ are required. These values can all be calculated directly from a trials database.

The first step is to select predictor and response variables. In the model specification stage, predictor and response variables need to be selected from databases or knowledge bases. In the implementation stage, predictor variables are drawn from spatial data, allowing the response variables to be spatially interpolated. Predictor variables may be biophysical or socio-economic in nature, as long as a relationship is expected with the selected response variable. The response variable depends on information available in the trials database, such as adaptation, yield or quality.

The certainty associated with the posterior probability distribution is derived from the individual certainty values for the conditional probability distributions. In this manner maps can be created showing suitability values based on probability distributions (Fig. 1a) and certainty values (Fig. 1b) for a particular response over an area of interest. Suitability values are calculated as weighted averages of the probability distributions at each location, so that a summary of the probability distribution can be shown in a single map. Areas showing low certainty values can be interpreted as environments where very little data exists.

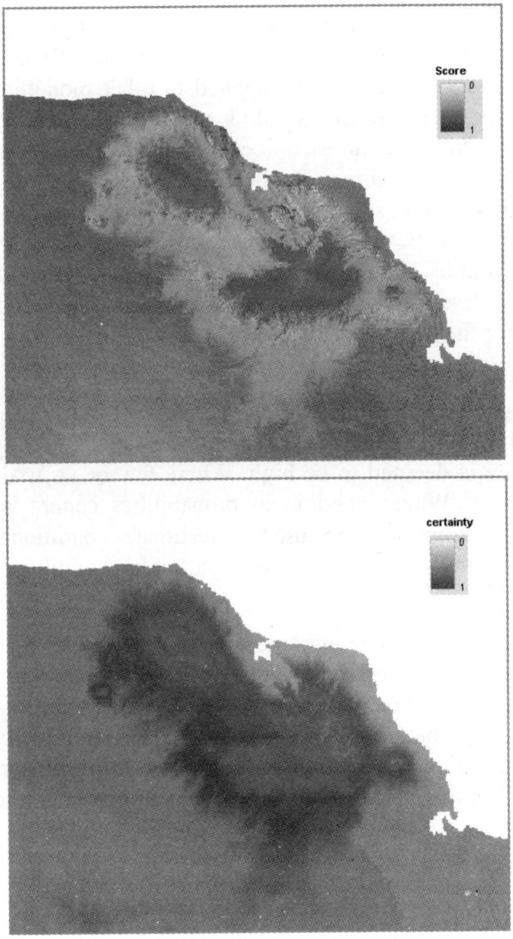

Fig. 1: Suitability (a) and certainty (b) maps

CaNaSTA also displays the information in graphs and tables, which can be manipulated by an expert user to update probability distributions with expert knowledge (Fig. 2).

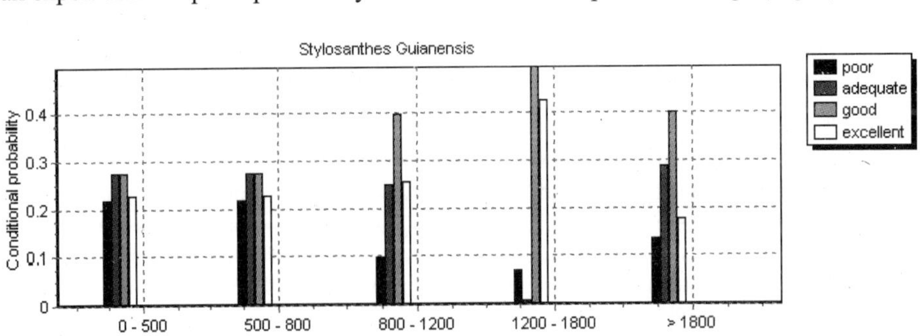

Fig. 2: Probability distributions in rainfall classes for the forage species *S. guianensis*

In addition, sensitivity values are calculated. This estimates the amount of change in the response when the predictors vary. For example a species with very specific niche requirements is likely to be sensitive to variations in the predictor variables, whereas a widely adapted species is likely to show low sensitivity.

There are various methods available to determine the sensitivity of the model to changes in states of the predictor variables. 'What-if' analysis analyses the change in outcomes if the state of one variable is changed (DITTMER and JENSEN, 1997). Sensitivity analysis also indicates which variables are important in a particular case.

Once locations have been identified where a particular response is likely, further analysis can be carried out to determine which predictor variables are important. These driving factors can either positive or negative, and can help with the analysis of specific conditions required for niche species or highly localised phenomena. Species can also be compared for suitability at specific locations.

4 Discussion and conclusions

CaNaSTA was initially developed as a tool to suggest niche forage species to smallholder farmers in the tropics (O'BRIEN et al., 2005). Recently, it has also been applied to coffee quality analysis (LÄDERACH et al., 2006), cowpea performance in tropical hillsides and carbon concentration in soils. The tool is proving useful for analysis of highly specialised crops, where little trial data may be available and variables influencing crop response are often crop-specific (ATZMANSTORFER et al., 2007).

Initial validation (O'BRIEN, unpublished data) indicates that CaNaSTA can provide useful results with very few trial data points (< 10), provided the trial data supply information for all response categories. Interpretation is enhanced by the provision of certainty maps. In addition, driving factor analysis allows exploration of spatially variable driving factors for niche crops. CaNaSTA is particularly valuable for examining potential spatial response, and the reasons thereof, of specialised niche crops.

Further research and development aims to provide additional analysis options for specialised crop response research, as well as the application of CaNaSTA to other diverse spatial research problems.

Acknowledgements

The research described in this paper was funded in part by the Bundesministerium für Wirtschaftliche Zusammenarbeit und Entwicklung (German Federal Ministry for Economic Cooperation and Development, BMZ). The author also acknowledges the support from the Deutsche Gesellschaft für Technische Zusammenarbeit (German Agency for Technical Cooperation, GTZ), the Diversification Agriculture Project Alliance (DAPA), the International Center for Tropical Agriculture (CIAT) and USAID.

References

ASPINALL, R, 1992, An Inductive Modelling Procedure Based on Bayes' Theorem for Analysis of Pattern in Spatial Data. International Journal of Geographical Information Systems, 6(2), pp. 105–121.

ATZMANSTORFER, K., OBERTHÜR, T., LÄDERACH, P., O'BRIEN, R., COLLET, L. and QUIÑONEZ, G., 2007, Probability Modelling to Reduce Decision Uncertainty in Environmental Niche Identification and Driving Factor Analysis: CaNaSTA Case Studies. In ZEIL, P. and KIENBERGER, S. (Eds.), Geoinformation for Development – Bridging the Divide Through Partnerships. Wichmann, Heidelberg, pp. 33–43.

BUSBY, J.R., 1991, BIOCLIM – A Bioclimatic Analysis and Prediction System. In MARGULES, C.R. and AUSTIN, M.P. (Eds.), Nature Conservation: Cost Effective Biological Surveys and Data Analysis. CSIRO, Melbourne, Australia, pp. 64–68.

CARPENTER, G., GILLISON, A.N. and WINTER, J., 1993, DOMAIN: A Flexible Modelling Procedure for Mapping Potential Distributions of Plants and Animals. Biodiversity and Conservation, 2, pp. 667–680.

CORNER R.J., HICKEY, R.J. and COOK, S.E., 2002, Knowledge Based Soil Attribute Mapping in GIS: The Expector Method. Transactions in GIS, 6(4), pp 383–402.

DITTMER, S.L. and JENSEN, F.V., 1997, Tools for Explanation in Bayesian Networks with Application to an Agricultural Problem. Proceedings of the First European Conference for Information Technology in Agriculture, Copenhagen, Denmark, June 15–18.

FAO (Food and Agriculture Organisation of the United Nations), 1978–1981, Report on the Agro-Ecological Zones Project. World Soils Resources Report 48, 1–4. FAO, Rome.

FERSON, S. and GINZBURG, L.R., 1996, Different Methods are Needed to Propogate Ignorance and Variability. Reliability Engineering and System Safety, 54, pp. 133–144.

GUISAN, A. and ZIMMERMAN, N., 2000, Predictive Habitat Distribution Models in Ecology. Ecological Modelling, 135(2–3), pp. 147–186.

HIJMANS, R.J., CONDORI, B., CARILLO, R. and KROPFF, M.J., 2003, A Quantitative and Constraint-Specific Method to Assess the Potential Impact of New Agricultural Technology: the Case of Frost Resistant Potato for the Altiplano (Peru and Bolivia). Agricultural Systems, 76(3), pp. 895–911.

HIRZEL, A., HAUSSER, J. and PERRIN, N., 2001, Biomapper 1.0. www.unil.ch/biomapper.

LÄDERACH, P., COLLET, L., OBERTHÜR, T. and POHLAN, J., 2006, Café especial y sus interacciones con factores de producción. Memoria del 2° Diplomado sobre Cafeticultura Sustentable, Tuxtla Guttierez Chiapas, Mexico.

MCCOWN, R.L., HOCHMAN, Z. and CARBERRY, P.S., 2002, Probing the Enigma of the Decision Support System for Farmers: Learning from Experience and from Theory. Agricultural Systems, 74, pp. 1–10.

O'BRIEN, R., PETERS, M., CORNER, R. and COOK, S., 2005, CaNaSTA – Crop Niche Selection for Tropical Agriculture, a Spatial Decision Support System. Proceedings of the Twentieth International Grasslands Congress, Dublin, Ireland.

O'BRIEN, R., 2004, Spatial Decision Support for Selecting Tropical Crops and Forages in Uncertain Environments. PhD thesis, Department of Spatial Sciences, Curtin University of Technology, Perth, 278 pp.

PEARL, J., 1990, Bayesian Decision Methods. In SHAFER, G. and PEARL, J. (Eds.), Readings in Uncertainty Reasoning. Morgan Kaufmann, San Mateo, CA, pp. 345–352.

ROWE, W.D., 1994, Understanding Uncertainty. Risk Analysis, 14(5), pp. 743–750.

PARADIS – A GIS Tool for Humanitarian Demining

Sébastien DELHAY, Vinciane LACROIX and Mahamadou IDRISSA

Summary

Landmines and UXO (UneXploded Ordnance[1]) represent a constant and long-lasting threat to the life of millions of individuals. Moreover, the resulting loss of agricultural fields, the inaccessibility of transportation roads or of water points due to these weapons have a strong socio-economic impact on contaminated countries. Demining is a critical issue since the clearance of these often-large contaminated areas requires investing much time and money. In this scope, PARADIS[2] aims at efficiently answering the needs of campaign actors such as the campaign planner and the field operators. The tool is built upon GIS technology and uses satellite imagery as a substitute for background maps, in order to represent all data involved in demining in their most explicit form: a map.

1 Introduction

Demining campaigns represent a huge amount of data: accurate perimeter of contaminated zones, locations of UXO's, villages, roads, etc. The planner's decisions depend on the reliability of those data: if only corrupted data are available to the planner, he will take bad decisions. For example, he could send out people to a minefield that has already been cleared because the data stating that it has been cleared were lost. The data gathered in the field must be safely conveyed to the Mine Action Center and stored in a central place in order to guarantee data integrity and avoid data loss; they must also be represented in their most explicit form so that the user can easily understand them and take pertinent decisions. Moreover, while maps are key elements for the management of demining activities, as demining campaigns generally take place in developing countries, there is often a lack of accurate spatial information for the region concerned. Satellite images may be proposed as a solution to this problem. An efficient GIS Mine Action tool should also be adapted to the users' specific needs and allow the planner to follow the evolution of the campaign from beginning to end, browsing through the identified scales of work, from global (country) to local (field) scale. These requirements are the development guidelines for the PARADIS system.

Mine Action information systems such as IMSMA typically propose one interface to enter the data (merely in their textual form) and another secondary interface – the GIS – for viewing the data on a map. Hence, the user has to switch from one interface to another in order to enter geographical data (e.g. a minefield) and its associated metadata (e.g. the minefield status). PARADIS rather considers the GIS as the central element of the system; hence, all the data can be entered from one single interface – the GIS.

---

1 "Munitions that have been used but which have not exploded as intended" (IMSMA).

2 PARADIS (a Prototype for Assisting Rational Activities in Demining using Images from Satellites) is nowadays funded by the Belgian Defense, and is developed for the Belgian Bomb Disposal Unit (SEDEE-DOVO).

Moreover, PARADIS aims at filling certain gaps in existing information systems by proposing specific tools. Indeed, in order to satisfy as many users as possible, the quoted information systems mainly propose tools that fit the common needs of most users, without integrating those more specific ways of working that vary from one demining organization to another. On the other hand, PARADIS proposes a modular approach that enables each user to have his own specific needs integrated into the interface. One may then see PARADIS as a tool complementary to other information systems.

2 PARADIS: a tool for mine action

The aim of the PARADIS project is to improve the planning of Humanitarian Demining campaigns using Remote Sensing data and GIS technology. The system addresses the different problems discussed in the introduction. In order to prevent data loss or corruption, in the planner's office a geographic database (or GeoDb) centralizes the whole set of data related to the demining campaign; in the field, a GIS interface provides a secure channel to convey data to the GeoDb. At each working step, the system guarantees the integrity of the data.

Moreover, to solve the problem of lacking maps, high- and very high-resolution satellite images and their interpretation constitute the system's background data and provide accurate and up-to-date maps of the region of interest to the user. This topic has been fully addressed in DELHAY (2003). Finally, in order to fit the needs of the users, the system is built upon four scales (global, regional, local and advancement) and actually follows the tasks assigned to the deminers from global to local scale; to each of these scales correspond dedicated tools, distributed in two separate – but complementary – interfaces: the Planning Interface designed for the campaign planner, and the Field Interface built for the field operator.

3 Designing two dedicated interfaces

The needs of the campaign manager differ from those of the field operator. While the manager needs a full-featured GIS interface to access the central database and take decisions at country and regional levels, the field operator needs a lightweight GIS interface to bring data with him and collect new information on the field. In order to reflect this duality, the PARADIS system consists of two separate interfaces. The Planning Interface runs at the central office on a standard PC or laptop, while the Field Interface runs on a PDA (Personal Digital Assistant).

4 Demining tasks

The work of the demining organization is composed of several tasks, which include:

4.1 Road reconnaissance

The field operator may ask the interface to take GPS points and store them as a new road. He may then enter the name of the road, the state during the dry and wet seasons, the presence of ammunition along the road etc. The planner is able to maintain an up-to-date road map for the region of work. In addition, he knows exactly which road is practicable during dry and wet season.

4.2 UXO reporting

Whenever the field operator finds a UXO in the field, he can enter its position, model, colour, markings, etc through the interface. He may also take a picture of the ammunition with the pocket camera that fits on the PDA. The planner has an overall picture of the location of the UXO's. He can visualize the places where there are most ammunition of a certain type in the whole region, rank districts by the quantity of ammunition found, etc.

4.3 Team management

Using PARADIS, the field operator is able to manage the team's personnel and stocks. For instance, when a metal detector breaks down in the field, the team leader will decrease the number of detectors of one unit. He will also enter the amount of explosives the team used to destroy the UXO's found, the volume of gas remaining in the vehicle tank, etc. The planner has a clear picture of the remaining stocks for each team. He is now able to place appropriate orders to keep teams' supplies up.

4.4 Contaminated area management

In the field, the deminers delineate the perimeter of the zone to be cleared. Then the PARADIS user enters the perimeter through the interface, and creates a grid on the zone: each cell in the grid corresponds to a square meter for clearance on the ground. As work advances, the user clicks the cleared cells and enters the possible ammunition found in those cells. The planner may follow the evolution of the work in each of the zones that are being cleared. If he finds that, the work of team A is slower than that of team B he may find it useful to transfer some personnel from team A to team B. He may also ask the system for an approximation of the time left to work on each zone.

5 Conclusions and future prospects

The two interfaces are still in constant evolution, based on the ideas and methods of work of the end-users and on feedback from the field. The system is, at time of writing, intensively tested in Belgium by the teams of the SEDEE-DOVO. They will also validate the system in operations abroad. In addition, different modules will be added to the system in order to fit the specific needs of other end-users of the Mine Action Community.

References

"A Guide to Mine Action", Geneva International Center for Humanitarian Demining. IMSMA website: http://www.imsma.org.

DELHAY, S., LACROIX, V. and IDRISSA, M., 2003, PARADIS: A GIS Tool for the Management of Humanitarian Demining Campaigns. Brussels, Belgium: International Conference on Requirements and Technologies for the Detection, Removal and Neutralization of Landmines and UXO (EUDEM2-SCOT).

DELHAY, S., LACROIX, V. and IDRISSA, M., 2005, PARADIS: GIS Tools for Humanitarian Demining. Brussels, Belgium: Information Systems for Crisis Response and Management Conference (ISCRAM 2005).

Application of the Fuzzy Logic Method for Landslide Susceptibility Mapping, as Feasible – Ready to Use Approach, in the Context of Development, Rio Blanco, Nicaragua

Harald SCHERNTHANNER

Summary

In June 2004, the Nicaraguan district of Rio Blanco was severely affected by a huge number of landslides, triggered by heavy rainfalls. Over 700 movements of changing size and type posed a great impact on the vulnerability of the regions infrastructure, homesteads and their inhabitants. The region can be described as underdeveloped and the authorities in the region need an instrument to conduct a reliable landslide susceptibility analysis. By means of a GIS, a fast and feasible analysis of the susceptibility of landslides can be done, as shown in this paper.

A GIS based fuzzy logic approach was used to develop a fast and feasible landslide susceptibility analysis of the region. The paper introduces to the context why landslide hazard assessment is relevant for development. After the characterization of the study area, different methods usable for a landslide susceptibility analyses are discussed and the fuzzy logic method is demonstrated in detail. The result of the susceptibility analysis is a susceptibility map, showing different zones of landslide susceptibility, from low to very high susceptibility. Urgent, medium to long-term recommendations for the region based on the result of the analysis show, that a landslide susceptibility analysis can be used as a good regional planning tool in developing regions.

1 Introduction

Despite the fact that developing countries are vulnerable to landslide hazard, they generally lack policies for hazard assessment and mitigation. For many applications, the most urgent problems are due to the lack of good quality data, on which to base a hazard assessment, in addition to insufficient funds and political will. Developing countries have an urgent need for ready to use, feasible, low cost methodologies in the field of landslide hazard analysis. GIS methods create possibilities to simplify landslide hazard analysis.

Influenced by landslides in Rio Blanco, Nicaragua, the fuzzy logic (fuzzy gamma) approach to analyse the landslide susceptibility of a region was applied, as example for a method that is flexible, fast and easy to use and can be based on little data of all statistic measurement scales. The analysis was undertaken in collaboration with INETER Nicaragua, Instituto Nicaragüense de Estudios Territoriales.

2 Landslide hazard assessment for development

It can be stated that a huge number of a relatively small to medium sized slope movements impose a great cost to human society, especially in developing countries. The economic and technological growth in developing countries causes a higher vulnerability to landslide hazard impacts. Further development often leads to an opening up of previously underdeveloped regions. This increasing human activity in former unpopulated areas additionally increases the vulnerability of humans to natural hazards. In developing countries, the perception of the landslide hazard is generally imperfect and is often attributed to a kind of higher authority. A consequence of this lacking perception of the landslide hazard in developing countries is the high vulnerability of humans and infrastructure, as mentioned before. Even minor and relatively slow moving slope failures can have a high cost, because of the destruction caused to infrastructures (JONES, 1992).

In developing countries, a need to increase the awareness for the issue slope movements exists. Landslide management (done by GIS means) should be integrated as an element to development planning. Instead of the traditional cost intensive response to the threat of landsliding, anticipatory planning and prediction of landslides in developing countries would lead to an avoidance of many disasters triggered by landslide (JONES, 1992).

3 Study area, "Rio Blanco", Nicaragua

The district of Rio Blanco in Nicaragua, Central America, was severely affected by the tropical storm number 14 (Onda tropical 14), in June 2004. A huge number of persons and homesteads felt victim to this tropical weather phenomenon, because of landslides and flooding. Over 700 different types of landslides showed clearly how vulnerable this developing region is to natural disasters. The 'municipio'-district of Rio Blanco is part of the department of Matagalpa. The capital of the district of Rio Blanco, also named Rio Blanco, is situated 110 km to the east of the city of Matagalpa and 250 km to the north-east of Managua, the capital of Nicaragua.

Rio Blanco counts a population of 33 195 persons. 9 254 persons live in urban area, 23 950 inhabitants live in the rural area. The population density is 47 people per km^2. The economy in the area is principally based on agriculture and cattle raising (INETER, 2005). As the region's climate poses a great impact for landslides, here a short description is given. The climate of Rio Blanco can be characterized as tropical mountainous, with over 9 rainy months per year. The average temperature varies between 20 and 26 degrees Celsius and the annual average precipitation varies between 2 400 and 2 600 mm per year. The highest values of perception are found in the months from June to August, and they are concentrated from the centre of Rio Blanco northwards.

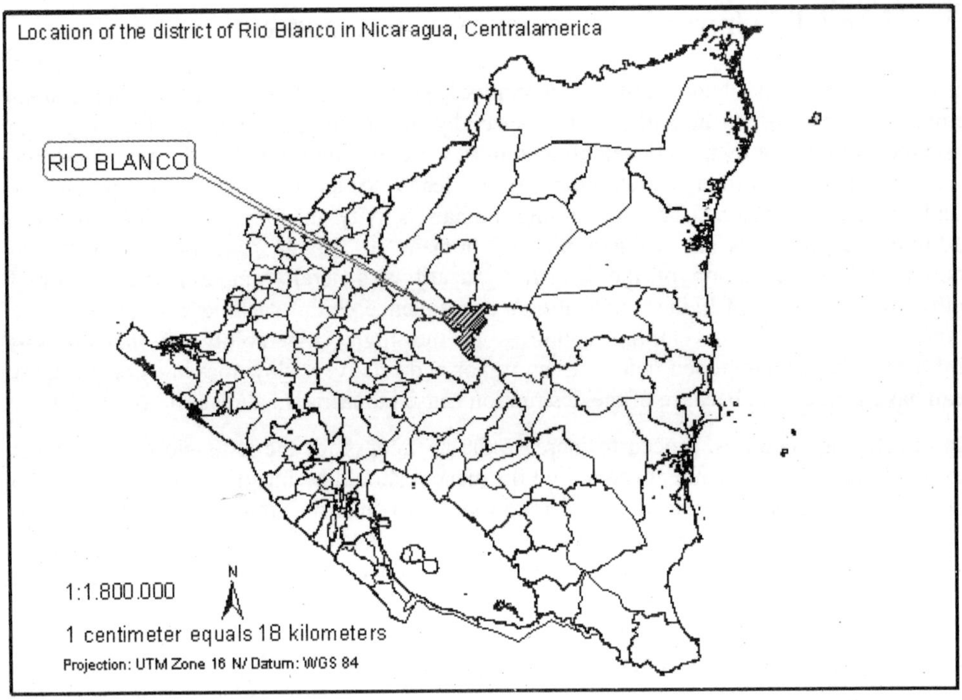

Fig. 1: Location of the study area

Rio Blanco counts a population of 33,195 persons. 9,254 persons live in urban area, 23,950 inhabitants live in the rural area. The population density is 47 people per km^2. The economy in the area is principally based on agriculture and cattle raising (INETER, 2005). As the region's climate poses a great impact for landslides, here a short description is given. The climate of Rio Blanco can be characterized as tropical mountainous, with over 9 rainy months per year. The average temperature varies between 20 and 26 degrees Celsius and the annual average precipitation varies between 2,400 and 2,600 mm per year. The highest values of perception are found in the months from June to August, and they are concentrated from the centre of Rio Blanco northwards.

As the agriculture is the most important factor of the economy in Rio Blanco and impacts on the landslides in Rio Blanco, here a description of the agriculture in the region is given: According to the agro-socioeconomic zonation of Nicaragua, the district of Rio Blanco belongs almost completely to the 'old cattle raising frontier of Nicaragua' (RUIZ and LOPEZ, 2005). The zone mentioned above can be characterized by a certain grade of diversification in its agricultural production, although the production of basic grains (corn, red beans) and cattle ranching are dominant. In the whole zone, 58% of the land is used as cultivated pasture. 10% of the land is actively used for forestry. Permanent cultivation only occupies 3% of the area, and bushes 19% (RUIZ and LOPEZ, 2005).

4 Characterisation of the slope instabilities in Rio Blanco

For generalization purposes the term landslide (or slope instability), will be used in this paper in the same sense of meaning, regardless of the characteristics of a mass movement. In total, including landslides before the "disaster" of June 04, 785 instabilities, different in size and form, have been counted. From 131 landslides, the coordinates have been taken. 118 of them lie exactly within the limits of Rio Blanco. These 118 instabilities were used for further analysis. As a form of generalization, the terms landslide and slope instability will be used in the same sense of meaning.

Table 1: Type of slope instabilities in Rio Blanco

Landslide type	Number of movements	%	Estimated volume of material moved (m^3)
Debris flows	483	61.5	50–6 000 000
Mud flows	227	28.9	20–6000
Rotational slides	71	9.0	200–265 000
Planar Slides	3	0.4	3000–300 000
Debris avalanches	1	0.1	> 1 000 000
Total	785	99.9	

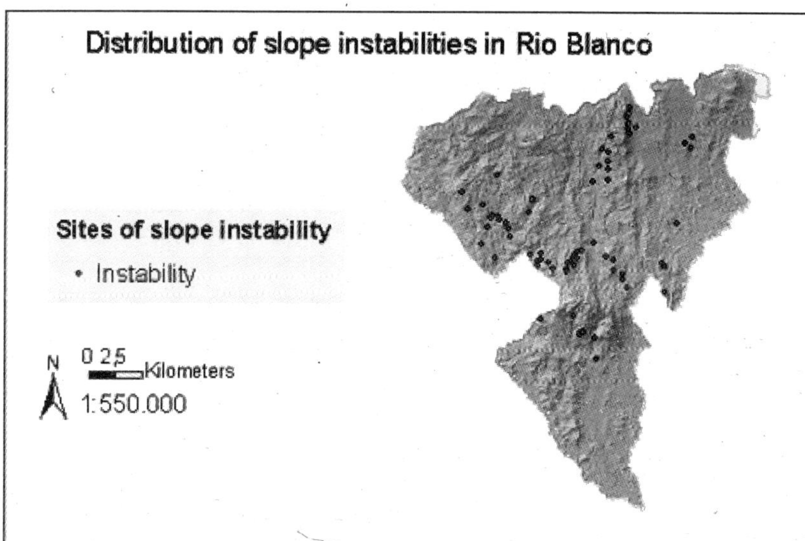

Fig. 2: Distribution of the 118 slope instabilities with available coordinates

5 Methods of Geographic Information Systems in mapping landslide susceptibility

As some confusion exists in the use of the term susceptibility, the meaning of the term has to be clearly defined. Researchers like BRABB (1984), VARNES (1984) used the following definition of the term that today still is the most common one in literature about slope failures assessment.

Susceptibility means the probability of the occurrence of a phenomenon, leaving out the factor time. Susceptibility is determined by the spatial distribution of already known phenomena and the correlation of the factors causing a phenomenon (BRABB, 1984).

Natural Hazard on the other hand, means the probability of an occurrence within a specified period of time, within a given area, of a potentially damaging phenomenon (VARNES, 1984).

Susceptibility maps predict where slope failures are most likely to occur and leave out the factor 'when' of the hazard maps (BRABB, 1984).

Hazard maps ideally show the location and the actual potential slope failure, and further provide information on the time and the probability of their future occurrence, their return period (BRABB, 1984). The hazard zone, extent of hazardous event and the internal division of an hazard, is shown together with a hazard rating, reflecting the magnitude frequency of an hazardous event. Such maps should be based on the evidence of past failures together with judgments as the likelihood of future events (JONES, 1992).

One of the main arguments why to conduct a landslide susceptibility analysis by GIS means is the *feasibility*. Feasibility means that a methodology has to be simple, inexpensive, should be ready for application when the need arises, and should be ready to be applied at a larger scale. Traditional landslide susceptibility mapping operates on a small scale. These *local methods* allow for very detailed investigations on single slopes. Normally these methods are not very useful when extrapolation and prediction for large areas is required. Such a prediction of can be carried out by *regional indirect methods* based on GIS. Regional methods allow the assessment of slope instabilities over wider areas on a medium to small scale (1:25 000–1:100 000).

Generally speaking, indirect methods consist in a series of overlays of landslide factor maps. Each of these maps displays the varying strength of a factor causing a landslide.

6 Fuzzy logic approach as example of a feasible, rapid method

Fuzzy logic works with the concept of partial truth. In the classical theory, membership is defined as 1 = true or 0 = false. It does not allow thresholds, a slope is susceptible for landslides (1) or not (0). In a fuzzy set, membership is expressed on a continuous scale from 1= full to 0 = full non-membership. Fuzzy logic applied on the search of slopes susceptible to landslides allows the use of thresholds. A slope might become susceptible at a steepness of 5% as lower threshold and can reach up to 30% as highest threshold.

Applying the fuzzy logic approach allows us to define the "steep slopes" in degrees of membership. (LORUP, 1999)

In landslide susceptibility mapping fuzzy logic defines the instability factors as members of sets reaching from 1, expressing the highest susceptibility, to 0, expressing no susceptibility of landsliding, therefore allowing different degrees of membership. The classes can be associated with fuzzy membership values in a table. The statistic category of the classes can be nominal, ordinal or interval. In a subjective judgment, values are chosen, to show the degree of a membership (BONHAM-CARTER, 1994).

Several fuzzy operators allow to combine the input maps (instability factor maps) in a series of steps:

- FUZZY AND
- FUZZY OR
- FUZZY algebraic product
- FUZZY algebraic sum
- FUZZY Gamma operator.

The input maps can be combined in a series of steps using different fuzzy operators. For example, instead of linking together all input maps ad once, sometimes it is more appropriate to link them together in intermediate steps by using different fuzzy operators, and finally these intermediate steps can be linked together to achieve a final result.

6.1 Fuzzy gamma approach

This method is used for the combination of input factor maps – a map that contains the factors causing slope failure. The fuzzy gamma approach is a compromise between the increasing tendencies of the fuzzy algebraic sum and the decreasing effect of the fuzzy algebraic product – other fuzzy operators to combine input factor maps. Advantages of the fuzzy gamma approach are the independence from the statistic level of the input data, further different scenarios can be compared from the maps produced during the process easily, and trends of the final results can be visualized.

7 Application of the fuzzy gamma approach in Rio Blanco, Nicaragua

7.1 GIS data acquisition in developing countries

Especially in developing countries, the acquisition of these data is very costly and time consuming, if not even impossible.

Based on GIS methods as the fuzzy logic approach, a reliable landslide susceptibility analysis in developing countries can be done in a cost- and time-effective manner, with a minimum of input data. For this analysis, *only* the following data layers have been used: the digital elevation model for the morphological features, lithology – geological formations, land cover (LANDSAT classification) and the soil types.

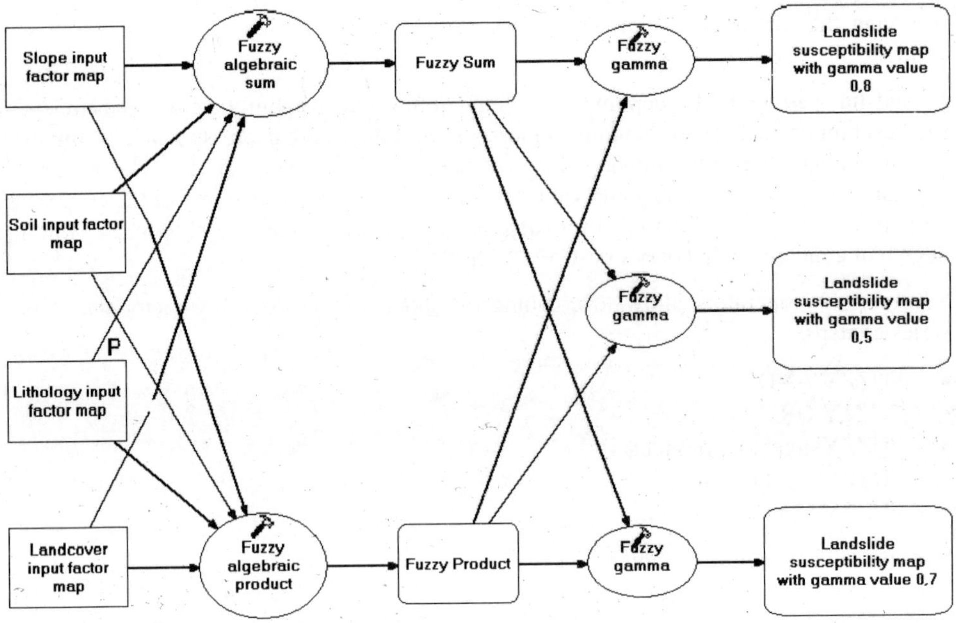

Fig. 3: Model builder diagram, combination of landslide instability factor maps, by applying the fuzzy gamma approach

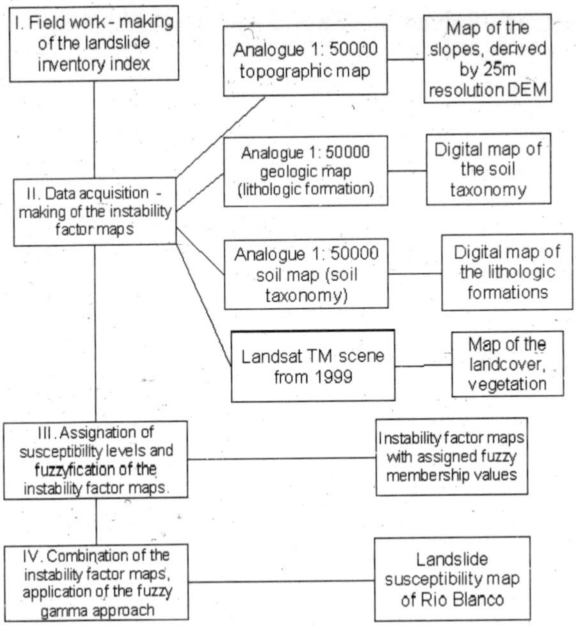

Fig. 4: Flow diagram for making a landslide susceptibility map

7.2 Scale of the analysis

The analysis concerns a regional, medium scale. The highest possible scale for a meaningful interpretation of the result of susceptibility analysis is 1:50 000, as 1:50 000 has been the scale of the base maps, from which the model-input data were derived.

7.3 Fuzzyfication and combination of the instability factor maps

As modelling approach, the fuzzy gamma approach was selected, as discussed in point 3. The membership values of the final susceptibility map reach from 0–0.2 (fuzzy membership value) showing the areas with the lowest landslide susceptibility until 0.7 to 1, reflecting the highest landslide susceptibility.

7.4 Urgent and medium to long-term recommendations based on the result of the susceptibility analysis

Urgent recommendations:

- Resettling of houses in zones of highest risk, to safe areas.
- Informing the population and education in the field of risk management.
- Development of an emergency and evacuation plan for the event of heavy rainfall, the main triggering factor of landslides in the area (see chapter 4).
- Distribution of information folders about prevention and recommendations what to do in the case of a disastrous event.
- Reinstalling and installing of new meteorological stations.
- Localization and the installation of signs of alternative ways of evacuation, usable in case of a disaster.

Medium to long-term recommendations:

- Development of a plan for proper soil management, management of forests, administration of water resources, reforestation, protection of natural resources and the environment.
- Development and support of land use planning.
- Control and reduction of cattle grazing inside landslide areas and the promotion of the use of hay to avoid too much pressure and weight on instable slopes.

8 Conclusion

By using the indirect fuzzy logic approach, landslide susceptibility mapping can be done under the circumstance of having few existing data about the factors causing the landslides.

The output susceptibility map of such an analysis can be used as a good regional planning tool but is not recommended for individual site-specific evaluation of the landslide susceptibility. One of the most important advantages of the approach is the inexpensive fast application by combining few available information.

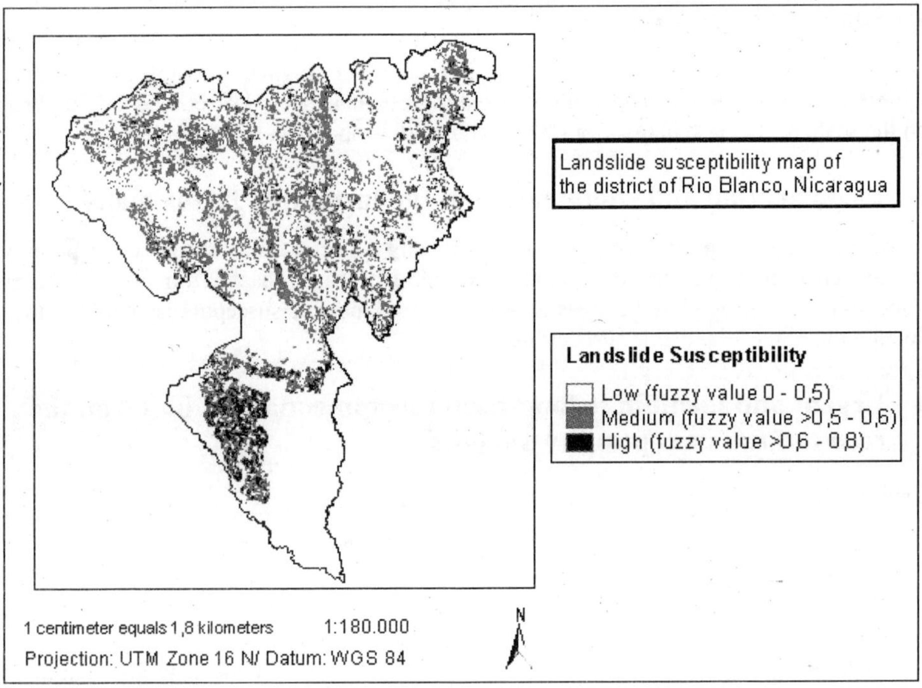

Fig. 5: Landslide susceptibility map

References

BONHAM-CARTER, G.F., 1994, Geographic Information Systems for Geoscientists, Modelling with GIS.

BRABB, E., 1984, Innovative Approaches to Landslide Hazard and Risk Mapping. Proceedings of the 4th International Symposium on Landslides, Toronto, 1, pp. 307–323.

RUIZ, G., LOPEZ, Y.M. et al., 2005, Revisitando el Agro Nicaragüense Tipología de los sistemas de producción y zonificación agro-socioeconómica, NITLAPAN, UCA.

INETER, 2005, Informe Prelimenar, Georiesgos en el municipio Rio Blanco, Managua.

INETER, 2004, Atlas Climático de Nicaragua, Managua.

JONES, D.K.C., 1992, Landslide hazard assessment in the context of development, In CALL, G.J.H, LAMING, D.J.C. and SCOTT, S.C. (Eds.), GEOHAZARDS, Natural and man-made. pp. 117–139.

LORUP, E., 1999, IDRISI WWW TUTORIAL. www.sbg.ac.at/geo/idrisi/wwwtutor/s_tools8.htm#fuzzy.

VARNES D.J., 1984, Landslide Hazard Zonation: A Review of Principles and Practice. United Nations Educational, Scientific and Cultural Organization. UNESCO Press, Paris, pp. 63–66.

Chapter 3

Institutional Requirements

The institutions of society, governmental and non-governmental, direct the path and pace of development. In an ideal situation, the spatial information they require for planning interventions and for monitoring progress is freely accessible to all, in the same accuracy and extent. The key challenge for implementing information systems is assuring that all actors can participate through provision of data and the extraction of information as required. However, information can only start to flow freely if a broad consensus about decision-making and participation exists in a society. Is change initiated by the establishment of information sharing platforms or have institutions to be conditioned first for the ability to share?

GIS for public participation: The case study on Ecuador elaborates on the change process in development planning and draws the attention to factors such as understanding the context, updating information, building scenarios, evaluating options and the effect of spatial visualisation on consensus building.

Development cooperation: A review of several projects for natural resource management in Malaysia, Indonesia, and the Mekong Delta, reveal some key issues for successful implementation. Systems need to work within a weak general communication and technical infrastructure, and initially implementers do have to cope with a lack of transparency and vested interests in natural resource management, but should actively address institutional and human resource development to ascertain the sustainability of the systems once the project comes to an end.

Early Warning: Providing information for action is the key objective of the Food Early Warning Service developed under the GMES Initiative of the EU. The service feeds information into existing regional, national and international early warning systems that monitor agriculture for food security, by continuously monitoring indicators such as Dry Matter Productivity or Vegetation Productivity. The resolution of the information is adjusted for regional, national and local level applications.

Capacity building: In Nigeria, a similar service was established at a local university, with freeware and free data, and disseminates striking evidence on the impact of urbanisation and the link between poverty and ecosystems integrity.

Meta information: *Who is doing what with how much money and where* is the central question the information system for development cooperation intends to answer. First developed for Mozambique, ODAMoz and ODAMap support donor coordination and efficient national development planning.

The problems are wide spread and well known: weak institutions cannot enforce protection measures; a top-down approach favours policy-driven rather than informed decisions based on a broad consensus; lack of transparency.

The audience voiced two main objectives to be pursued:

- Strengthen the capacity of universities as a main actor for change and as the provider of a dissemination platform open for all;
- For 'really making a difference', projects/interventions have to be planed and implemented over a longer time period to allow for partnerships to grow and perceptions to change.

GI for Development – Can GIS Challenge Existing Power Structures? Working Experiences from Ecuador

Richard RESL

Summary

Shortcomings of democratic, well informed, collaboratively with the public processed, inter-institutional and integrative local and regional planning procedures, are increasingly observed to provoke a counterproductive perception, procedure, trend and outcome for the development in Ecuador.

Interestingly enough, there are sufficient topographic and thematic spatial data available, plenty of GIS tools implemented within the national and local governments, NGOs and students gathering information for each and every place in Ecuador, and there are at least very helpful basic methodologies of GIS implemented.

However, when it comes to design, implement and sustain consistent, locally driven, regionally integrated and 'globally' supported development strategies, these advanced spatial datasets, tools and methodologies are not ready to communicate essential problem scopes to all stakeholders, and even less to the public. Participation, consequently, is reduced to 'share what's there', but does not come up to its potential role and the citizens' or communities' expected benefits of individual and collective involvement, preparing a fertile, well at least fair ground for the often promoted inclusiveness paradigm within good governance designs for development regions.

The question of who is responsible for those shortcomings can be addressed in different ways:

Are the existing power structures responsible? Is it more an organizational and technological difficulty to be overcome? Or are the citizens of Ecuador not ready to make use of their rights and responsibilities regarding spatial problems and development within their environment?

To address the importance of GI, to find out about the dynamics behind the questions raised above and to overcome the mentioned shortcomings of a critical involvement of well and adequately informed stakeholders, a methodological framework for spatially enabled citizen-driven and expert supported information and planning platform with a solid structure is called for. Two different scenarios for ongoing development tasks, both in an urban and rural setting, are provided to pilot and discuss alternative system designs for participatory spatial planning and management, and examine the roles of mentorship for a process oriented guidance and capacity building on how to use spatial information and apply GIS methodology within a given context.

1 Introduction

Geographic Information (GI) works for development projects, both with a top down strategy for most public institutions and a bottom up strategy for some NGOs but mostly community based initiatives throughout Ecuador. Is it to plan, push, control or hinder a certain development at national, regional or local level that spatial information plays a key role in different project stages: from analyzing necessities within a region, requirements and limitations due to spatial characteristics, deciding both on thematic and spatial priorities, up to monitoring and evaluating the spatial implications of processes and outcomes.

2 The problem scope

Can GIS represent the many and convince the few that development aimed at real common wealth is a worthwhile strategy towards sustainability? Sometimes that seemed to be not directly a question of convincing somebody, but more of putting pressure on decision makers merely by the capabilities of GIS, like inclusiveness, integrity, certain objectivity, transparency, participatory and collaborative contents, accessibility and worldwide authority. Or will GIS in hands of who so ever replicate the dominating power structures, just with that participative touch?

The paper discusses critical GIS design and implementation issues in a participatory setting of two apparently contrasting important development tasks in Ecuador:

- Infrastructure Development for the new millennium in QUITO – the new road and air transport system in the TUMBACO valley
- Natural resource development in the south eastern ORIENTE, Amazonian Lowlands of Ecuador – conservation, mining and oil business on indigenous lands

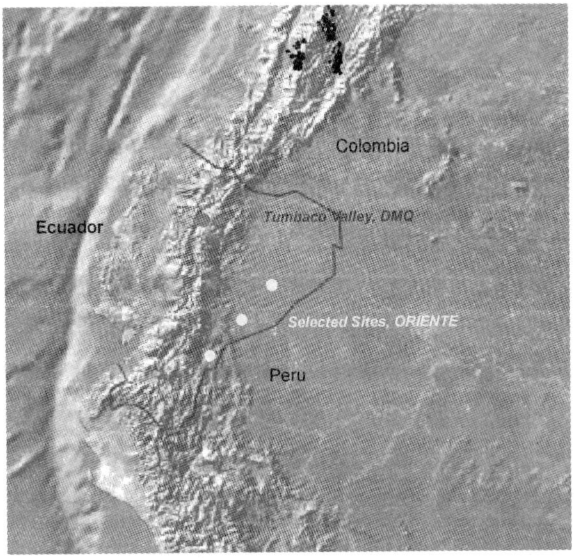

Fig. 1: Study Areas

For both study areas, it holds true that the local population experiences a strong pressure on increasingly scarce resources, especially on land and water, water and air quality, areas with a fair healthy quality of life, natural vegetation and green belts. Those areas experience an ongoing subtle alienation of traditional culture and ancestral value system, but are lacking a self-determined transformation process based on adequate information, and feel the increased dependency on actors, policies, resources and decisions from outside the local community.

Table 1: What is demanded from the civilians vs. what have they experienced so far? How can GI address these demands, support a coordinated action plan and change the experiences of alien determination on one side and inabilities of constructive participation on the other side?

Demands	Experiences
Respect of their Andean and Amazonian cultures regarding their community based settlement, landuse and economical structures	Ignorance and power demonstrations of local, regional and national governments and economic powers
Participation in spatial planning	Invitation of "government friendly" representatives
Information	Insufficient, inadequate information, propaganda or disinformation
Minga ("tax by working")	Concessions to non-local firms or capital
Dialog	Mass media campaigns against "progress reluctant citizens"
Capacity building, own capital and administrative determination	Non-local design, capital, expropriation and cost recovery shifted onto generations of users and no users of too big dimensioned spatial infrastructure
Self determination	Alien determination
Citizen oriented landuse measures	Speculation
Living quality for the many	Isolated luxury in gated cities
Rights to accomplish basic needs	Dependency on public services

3 Methodological framework

GIS has been used as a decision making tool since its creation. Until recently the tool has been available to only a small group of GIS users and individuals with access to and knowledge of GIS. This is still somewhat the case in Ecuador and true for the study areas of this research, but could be overcome by facilitating a supportive environment for a spatially enabled citizen-driven and expert-supported information and planning platform which basically is developed around the three following methodologies:

Public Participation GIS – implemented as "GISciudadano" and "AmazonGIS"

GIS for Participation – implemented as "PLANIFICAMOS"

GI User Networks – implemented as "Minga Tecnica", "Asociacion de Profesionales Independientes API" and "AmazonGISnet"

Suitable system architectures intend to make both instruments and data available for a consistent and ongoing collective effort strengthening a collaborative approach to development issues for a broad profile of users, and center on a methodological framework composed of GIS tools in combination with a bundle of methods for interactive participation following two basic design criteria:

- Culturally and socially sensitive GIS tool interface, setting, data content, data structure and output design. Design features focus on language, map signatures and symbols, legends, visualizations exploring multimedia mapping.
- A network approach as the core organizational setting

Especially the above-mentioned network approach explores facilities, knowledge and capacities regarding system maintenance and management, thus lowering undesired dependencies. Open source software and public domain spatial data sets should lower direct costs, and allow for a worldwide cooperation regarding development of tools minimizing the vulnerability of the system. The cost and policy factor of system administration has to be carefully handled, but is not yet a decisive issue within the experimental phase. Alternatives will be elaborated within the institutional contexts of the study areas.

4 Analysis and results

Whereas the design work on structured system architecture started just recently in 2005, several years of experience working with essential components of the methodological framework can be analysed and yet discussed as preliminary results.

The first part of the analysis is dedicated to demonstrate concrete achievements of the system design of GISciudadano and Planificamos (Tumbaco Valley) within its first implementation phase.

The second part comments on general characteristics regarding the network approach used for both study areas.

The third part gives some examples from both study areas and evaluates briefly their influence on existing power structures.

4.1 System design GISciudadano and Planificamos – first implementation phase – study area "Tumbaco Valley"

Currently, this research work evaluates the outcomes of the first implementation phase and elaborates the settings for the second one.

For the Tumbaco Valley the following system designs are implemented:

- Institutional framework
- Data preparation for on- and off-line publication
- Spatial Data Management System GISciudadano and Planificamos
- Fieldwork within participatory settings

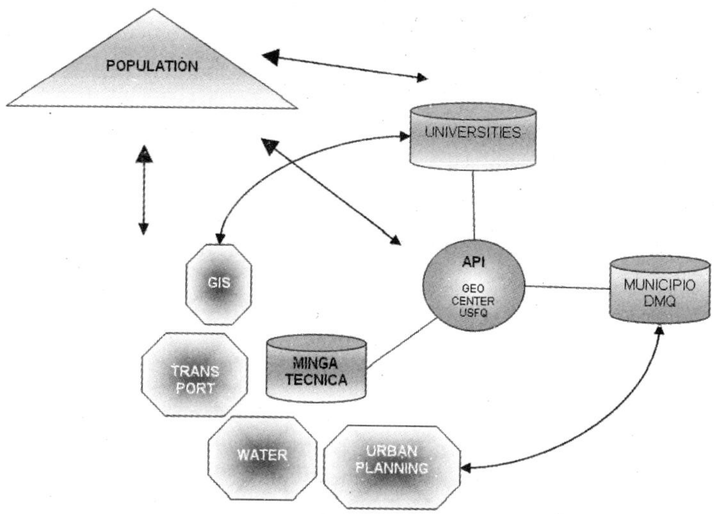

Fig. 2: Institutional framework GISciudadano, 1st implementation phase

4.2 Formation of GI user networks

During the last 5 years of participatory work on development issues in the two study areas, several individuals showed intensive interest in managing GI for their communities. Beyond the scope of informal and formal working groups, information sessions and typical community mapping scenarios both in the Tumbaco Valley and the Amazonian Lowlands, GI users asked for capacity building and continuous support. As a result of this motion, the author catalyzed the organizational setting for three different network initiatives.

AmazonGISnet started in 2001, and comprehends today a stable group of 25 members representing 9 indigenous nationalities and their respective organizations, situated mostly in Ecuador, with family organizations in Peru and another one in southern Colombia. One of the most valuable benefits emerged from the fact that a 'regional delegation', so to speak, came together without any political motive, and would start to treat regional aspects of spatial problems rather than on purely isolated local issues. Quickly, work focused on thematic issues and a standardized regional spatial database could be built at once.

Asociacion de Profesionales Independientes API, was also formed in 2001, and represents an expert group of citizens within the participatory institutional design. From the beginning, a group of 12 people were collaborating under the coordination of the author, with the personal and infrastructural support of the GEOCENTER of the University San Francisco de Quito USFQ and GEOinfo, Ecuador. Their contribution represents an essential professional backup on design and management tasks of the system, but can also serve as a role model for facilitation and mentorship, including capacity building. This concerns the application of GI-tools and techniques provided for the Minga Tecnica initiative for their participatory information and planning sessions, and as a driving force for public acceptance and awareness.

Minga Tecnica del Valle de Tumbaco, started just recently in 2005, as a result of many informal sessions with a dominantly informative character demanded by the population of the Tumbaco valley during the last 5 years. Three citizen workgroups gathered around thematic priorities identified as transport, water, urban planning. A fourth group showed interest to coordinate the citizens' work especially on GI management tasks.

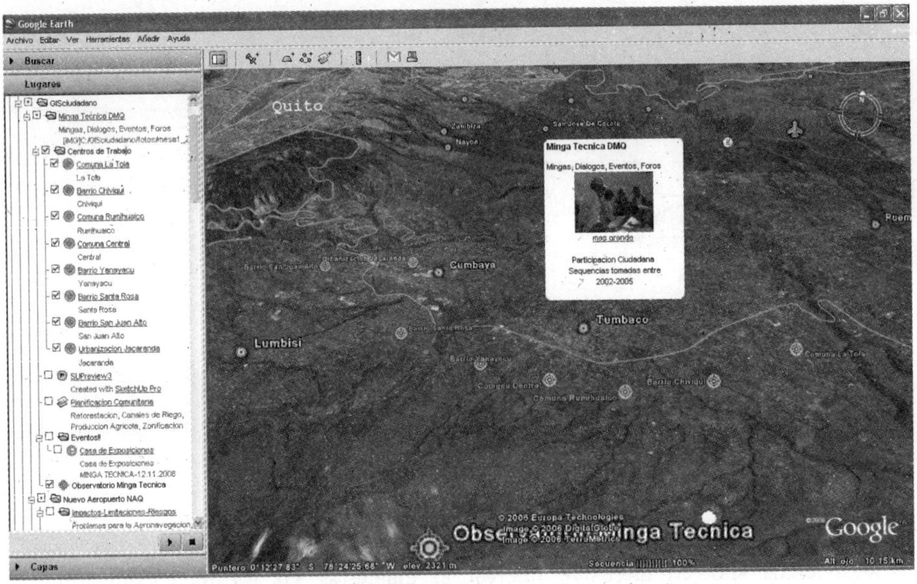

Fig. 3: Valle de Tumbaco, GISciudadano, 1st implementation phase; Comunidades, Minga Tecnica 2005, Google Earth 2006

4.3 Examples of working products and their challenging character on dominating power structure

What characterizes dominating power structures and their decision making mechanisms applied for or in the study areas?

- Dominant top down planning approach
- Confusion about interests, stakeholders and their competences
- Lack of integrated, consistent information
- Lack of spatially enabled and structured information
- Lack of up-to-date data
- Inadequate information flow and accessibility options
- Propaganda and disinformation
- Lack of capacity to manage, analyze and model spatial information
- Lack of organizational structure to handle the citizens' participation

Table 2: Example of API Network Initiative using GISciudadano and, Planificamos

Example of Products elaborated within "API"	Impacts created on existing power structures in charge of development
Evaluation of suggested access road infrastructure to New Airport Quito NAQ site and proposition of alternatives with less impact (presented to the City Council, Quito, 2003) and evaluation of technical feasibility of NAQ site (presented to the public at FORO CIUDADANO, Universidad San Francisco de Quito – USFQ 2005) *Fig. 4* shows one of the proposed bridge constructions connecting the Panamericana Road with the NAQ site, a connector road of just 4 km passing by the village of Oyacoto would be necessary to complete the functional and cost-effective access route. *Fig. 5* shows severe limitations of the proposed NAQ site for secure aeronavigation, due to climatic, topographic and geologic constraints.	Upon presentation to the respective authorities of urban planning and the City Council of Quito, Newspapers, Television, Radio, Professional Colleges, Universities, Communities and Citizen Groups during 2003, the original project proposal was rejected due to the arguments and recommendations. The proposed alternatives were taken into account for a new study phase contracted in 2004. A technical report presented at the FORO CIUDADANO in Nov 2005, organized by API, the "Observatorio Ambiental" and students of the USFQ, opened a public debate on limitations and potentials of the NAQ. Feasible alternatives were proposed.

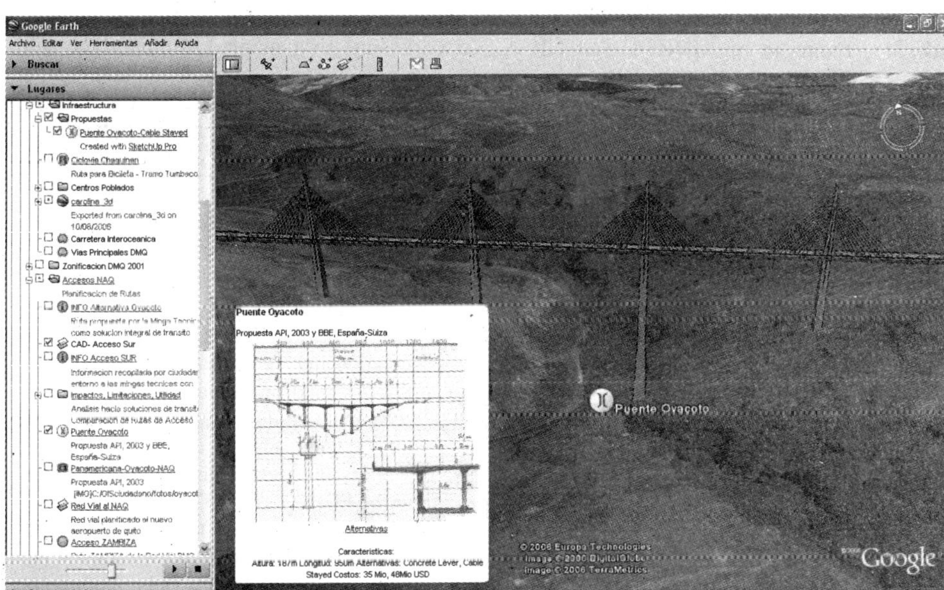

Fig. 4: ExampleAPI, GISciudadano, 1st implementation phase, Design of Oyacoto Bridge to NAQ site, API 2003, Google Earth 2006

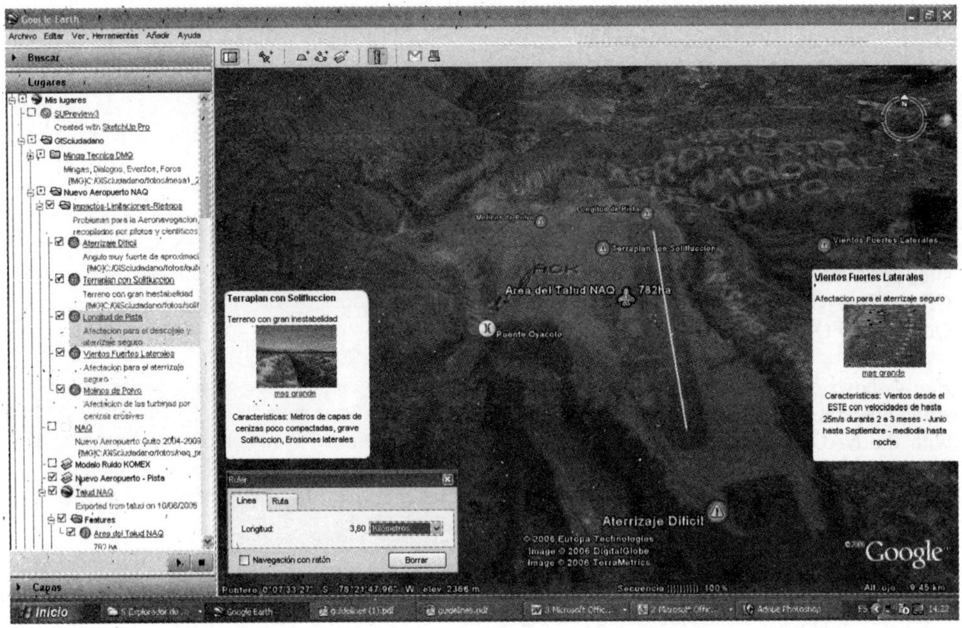

Fig. 5: Example API, GISciudadano, 1st implementation phase NAQ site limitations, API 2005, Google Earth 2006

Tab. 3: Example of MINGA TECNICA Network Initiative using GISciudadano and, Planificamos

Example of Products elaborated within "MINGA TECNICA"	Impacts created on existing power structures in charge of development
Evaluation of impacts of proposed road network for the Tumbaco valley (presented to the public at FORO CIUDADANO, USFQ 2005) and Design for inventory of irrigation water system for the Tumbaco valley (presented to the Directory of Users and technical board of governmental administration, CORSINOR 2004) *Fig. 6* shows the impact of the 30 m wide road (blue) on properties, homes, existing infrastructure, where the scenario could possibly look much worse if a corridor of 100 m broad (yellow) would have to be depopulated as mentioned in the original right of way plans from the 70ies. Meanwhile the area gets densely populated. *Fig. 7* shows the digitizing of additional feasible water channels upon request of users applying the citizens' planning tool PLANIFICAMOS based on the open source software gvSIG developed by Generalitat Valenciana, Transport Department, Valencia, Spain.	Representatives of MINGA TECNICA presented a summary of all community reunions and citizen workgroups and their observations on impact assessments, cost-benefit analysis of the delivered road network study. Further, additional components of alternative traffic solutions based on proper observations and technical support by professionals within and invited to the citizens' Transport workgroup, were submitted to the public at the FORO CIUDADANO in 2005. In order to maintain the irrigation water system administration in the hands of the users, strong arguments had to be presented to both user groups and the government's commission on water rights. A possible privatization to a service company is since then on hold.

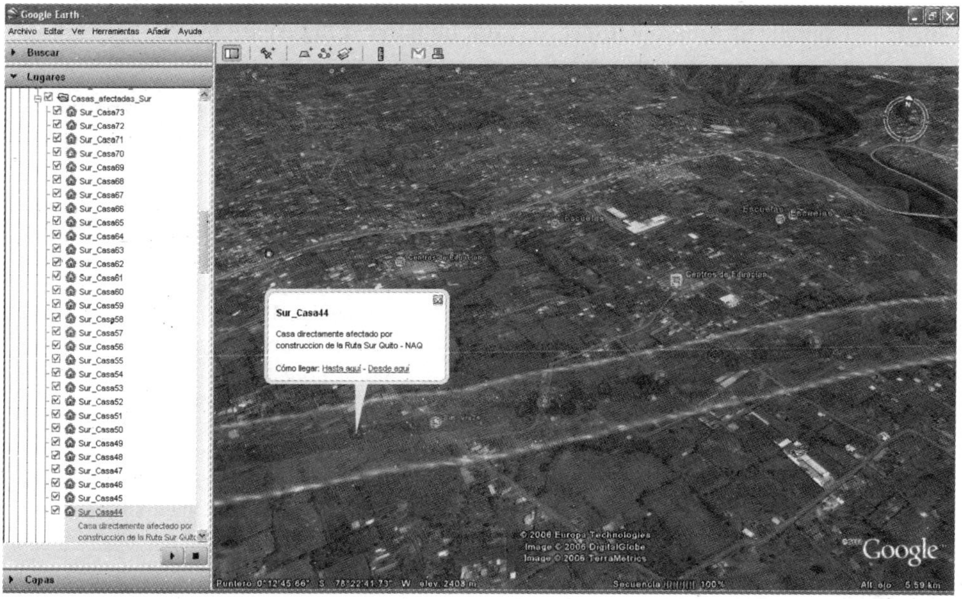

Fig. 6: Example MINGA TECNICA, GISciudadano, 1st implementation phase Construction Impact of Ruta Sur in Comuna La Tola, Minga Técnica 2005, Google Earth 2006

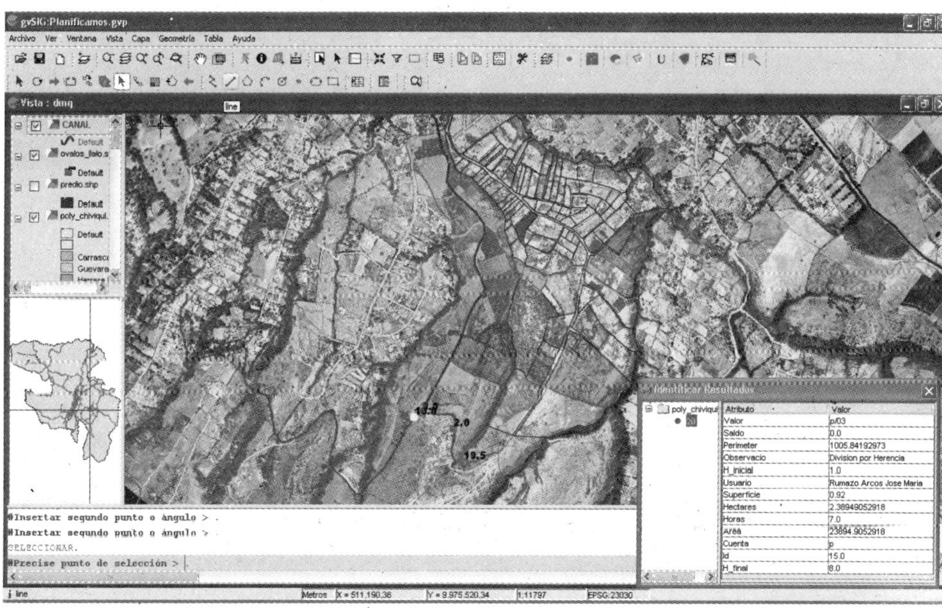

Fig. 7: Example MINGA TECNICA, GISciudadano, 1st implementation phase Irrigation Water Management Tumbaco, Minga Técnica 2005–2006, gvSIG, 2006

Tab. 4: Example of AmazonGISnet Network Initiative using AmazonGIS

Example of Products elaborated within "AmazonGISnet"	Impacts created on existing power structures in charge of development
Partial Demarcations and Legalizations of Indigenous Territories of the Shuar, Achuar, Shiwiar, Quitchua Nationalities in south eastern Ecuador (supported by Pachamama and IQBSS, Ecuador 1999–2006). *Fig. 8* shows the official map format for legalization of Asociacion Iniayua, Organization FIPSE, Shuar Territory, Transkutuku.	Legalization of Land stands probably for one of the most important demands within the participatory planning initiatives. The technical work done and submitted to the national authorities in the past 7 years include around 800 000 ha, 4 nationalities, and more than hundred communities. Demarcation and the elaboration of proper base maps and management plans would never have been possible without a strong participatory strand, making local people to GPS, GIS and RS technicians and the local communities to authors of their spatial determination.

Fig. 8: Example AmazonGISnet, Partial Demarcations and Legalizations of Indigenous Territories of the Shuar, Achuar, Shiwiar, Quitchua Nationalities in south eastern Ecuador Mapa Asociación Iniayua, Territorio FIPSE, 2006

Probably most important to notice is the process orientation and the fact that these products are not single submissions of a contracted work. They much more represent a continuous working process, and are periodically consultable with actualized information. Now that almost everyone's house and backyard is visible on Google Earth, and the "zoom in" factor is not of concern, a "zoom out" might do us well to get better connected.

The intangible values that, with time, could be translated into huge monetary benefits in terms of savings or locally sustained successful investments, are an important increase of

- information and awareness on all levels
- local identity and responsibility
- comprehension of spatial relations, dynamics and competences
- prepared ground for reflection and dialog
- support for orientation and decisions
- support for coordinated, collaborative and collective action
- image, self esteem and encouragement

5 Critical research issues

As a high degree of participation, over time and with a claim on successful vertical integration, is intended to be achieved, the system design consequently needs to consider particularly operating standards of GI. Components have to be developed continuously to meet the best knowledge and practice of tools and information that stand for

- Interoperability – The National GISplatform and data authority, ownership, inter-institutional design, public access
- Metadata management – GeoData Standards
- Open architecture
- Participatory planning – a multi-methodological toolbox for spontaneous participation supported by a well structured data repository.
- System support and capacity building – user networks and online resources
- Open source, public domain software for the citizens
- Online spatial and map server technologies – a powerful way to make development a citizens' decision
- Sustainability – GIS data and system maintenance beyond the project scope

6 Evaluation and conclusions

So far, it is clear that development issues require a highly complex and integrated systemic approach, and definitely do not influence current GIS designs and applications in Ecuador.

Will the implementation of a methodological framework for a spatially enabled citizen driven and expert supported information and planning platform make any difference for the study areas? If yes, could this framework be transferred and translated to other similar or even very different environments? As the research work is still under way, it seems appropriate to conclude with some critical questions that should somehow be evaluated regarding the proposed effects that do or do not come along.

- Does participatory work based on GI raise consciousness, problem awareness, civil responsibility and will it enable the citizens to act locally and regionally, individually and collectively on their behalves?
- Do collective processes of planning significantly support efficient resource management and sustainable spatial development?
- Does participation compete with the tasks of the planning authorities, or would a well structured vertical integration boost the efficiency and quality of the dominating planning and decision structures?
- Does participation have to be chaotic, temporary, selective, manipulated, intrinsically subjective and odd?

Learning lessons might take more time than often assumed. After all, the bigger challenge will be on how to apply both expert and local knowledge to assure that the collected information works for cooperative community planning initiatives towards a sustainable management of the environment.

Ongoing efforts throughout the next phases of design and implementation of this research initiative will be published at www.amazongisnet.edu.ec and www.gisciudadano.org.ec, which are currently under construction.

A View on Implementation Challenges of Geographic Information Systems in Natural Resource Management – Examples from South East Asia

Gernot RÜCKER, Georg BUCHHOLZ, Anja HOFFMANN, Christoph FELDKÖTTER, Isabelle KOLLAR and Anke CAMPHAUSEN

Summary

Spatial information derived through Geographic Information (GI) technologies has become an essential component in supporting decision making processes in Natural Resource Management. In this paper, we present experiences made during the implementation of Geographical Information Systems (GIS) in four international cooperation projects carried out in Malaysia, Indonesia and the Mekong Region. Based on these experiences, we draw conclusions for the requirements of design and implementation of GI systems in the framework of cooperation projects in natural resource management. During implementation of these systems, we encountered some obstacles (or challenges), which, we believe, are common in such projects. Among those count: a general reluctance to sharing of knowledge and data, lack of transparency and vested interests in natural resource management, a deficient general communications infrastructure, and the need for institutional and human resource development not at least to address IT/GIS capacities of stakeholders. Furthermore, the ability of organizations to maintain and expand systems after the international cooperation project ends are often limited. Setting up and implementing an efficient and sustainable GIS system is therefore not only a technical issue, but also poses demands on the organisational as well as communication culture of the involved organizations (both development organization and partner) and requires a comprehensive strategy that addresses the system development process as well as related institutional, political, and capacity building issues. In conclusion, we try to outline the main points that need to be considered when developing such a strategy.

1 Introduction

Over the last two decades, decision making in Natural Resources Management has become more and more complex, just as it has in any other field characterized by high levels of complexity and uncertainty. This is due to the increased amount and often conflicting information available as well as a more complex organisational landscape, in which new stakeholders from the civil society demand to be taken into consideration. In the South East Asian context – but also elsewhere in the world – decisions in management of forest and land resources have to be made in an environment characterized by increasing depletion of once abundant and seemingly inexhaustible goods. Therefore, natural resource managers, when attempting to decide in a factual and rational manner, have to base their decisions on large amounts of spatial and non spatial information and knowledge originating from various sources. Spatial information derived through Geographic Information (GI) technologies has become an essential component in supporting these decision making processes.

In this paper, we present experiences from the implementation of Geographical Information Systems (GIS) in four international co-operation projects carried out in Malaysia, Indonesia, and the Mekong Region:

1. A forestry information system developed for the Malaysian forest service,
2. A fire information system developed for the Indonesian province of East Kalimantan,
3. A fire early warning system developed for the Indonesian province of South Sumatra, and
4. A regional forest cover monitoring system established in cooperation with the Mekong River Commission.

Based on these experiences we draw conclusions regarding the requirements of design and implementation of GI systems in the framework of cooperation projects in natural resource management. One of our main theses is that implementation of GI systems in this context is at least as much a challenge in term of capacity building and institutional development as it is in the technical field. We also feel that this aspect is often underestimated by both the development organization and donor as well as by the partner organization(s).

2 The projects

2.1 Implementing a Forest Management Information System in Malaysia

Malaysia is one of the main exporters of tropical wood products. In the recent past, the provision of environmental services is coming more and more into focus of government policies within Malaysia. Sustainable Forest Management is the only way to balance the economic, environmental and social functions of forests for the benefit of the country and meet the increasing demand of civil society and consumer countries to environmentally friendly production of timber. To achieve these objectives forest managers in Peninsular Malaysia require detailed spatial and non-spatial information on logging areas such as forest functions, boundaries, topography, accessibility and the location of harvestable trees.

The Forest Management Information System developed within the Sustainable Forest Management and Conservation Project (SFMCP) funded by the German Federal Ministry of Economic Cooperation and Development is a combined GIS and data base application. It is designed to enable the State Forestry Departments to systematically and rapidly produce operational maps and reports such as boundary maps, forest inventory maps, timber harvesting maps and other maps, tables and reports in accordance with sustainable forestry requirements (RÜCKER and BUCHHOLZ, 2003). With support for planning and monitoring on compartment-level the implementation of Sustainable Forest Management is facilitated through an increase in data consistency, completeness, quality and a higher transparency of forest management processes.

2.2 A Fire Management Information System for East Kalimantan province, Indonesia

In Indonesia fires have devastated millions of hectares of forest and land during the last two decades. Sumatra and Kalimantan are the areas most affected by past fires from 1982 until 1997. In 1998 East Kalimantan alone was affected by fires covering about 5.2 million hectares of land and forest (SIEGERT et al., 2001). Most of the fires were related to land use change and natural resource exploitation; and over-logged forest were heavily affected (SIEGERT et al., 2001). Within the framework of the Integrated Forest Fire Management (IFFM) project, a technical cooperation project under a bilateral agreement between the governments of Indonesia and Germany, funded by the German Federal Ministry of Economic Cooperation and Development, an Integrated Forest Fire Management system in East Kalimantan has been established to reduce the negative impacts of fires. The concept of integrated fire management is based on two major elements: first, the involvement of local communities and second, a coordinated approach for developing a comprehensive fire management system. In East Kalimantan, this approach consisted of three functional units: fire information, fire prevention, and fire operations.

The objectives of the Fire Information System for East Kalimantan (FIS) was to compile and analyze fire relevant data to monitor vegetation fires, determine preparedness levels, identify fire risk areas, and assess the current and predicted fire situation (HOFFMANN, 2003). This information (maps etc.), necessary to make recommendations, are provided to political actors, organisations, and other land management agencies dealing with fire management to support planning, prevention, fire suppression and, in the end, an efficient fire management policy.

2.3 Developing Fire Threat Analysis for South Sumatra Province, Indonesia

The Indonesian province of South Sumatra is a fire prone area with a history of catastrophic peat and forest fires, mainly related to changes in land use (BOWEN et al., 2001, CHOKKALINGAM et al., 2005). In order to increase the institutional capacities needed to prevent and react to land fires, the EU sponsored "South Sumatra Forest Fire Management Project" supports the establishment of an early warning system, consisting of Fire Danger Rating (FDR) and Fire Threat Analysis (FTA). SSFFMP operates in a multi-stakeholder environment and cooperates with all agencies involved in land management and fire-related issues at province and district levels as well as with local communities and non-governmental organizations.

Fire Threat Analysis is a GIS-based approach that tries to map the overall threat by fires in an area. It consists of an assessment of the risk of ignition and the potential fire behaviour on the one hand, and the fire suppression capacities and the values at risk on the other hand.

The system consists of a knowledge-driven GIS application to automatically produce output maps on both probability and likely consequences of a fire starting at a certain location. The objective of the application is to provide decision makers with a basis for planning prevention campaigns, assessing suppression capacities and support fire-sensitive land use planning.

2.4 A Forest Cover Monitoring System for the Lower Mekong Basin

Forests throughout major parts of the Lower Mekong Basin (LMB), comprising portions of Cambodia, Laos, Thailand, and Viet Nam, are under severe threat from logging and conversion to other land uses. Deforestation has important environmental consequences as it may for instance cause watershed degradation with impacts on water quality and quantity. Consequently, the Mekong River Commission (MRC) in the early 1990s identified forestry issues as crucially important for the sustainable development of water and related resources in the region, and forest cover monitoring as an indispensable prerequisite to address these issues.

The German government provided related technical support from 1992 to 1999 through the Forest Cover Monitoring Project (FCMP). The FCMP supported the establishment of a regional GIS based natural resources information system including a forest cover monitoring and trend analysis system. Owing to very basic communication infrastructure in the LMB during the 1990s, the system was established as a distributed, decentralised system in various national line agencies and the Mekong River Commission itself. Consequently, the FCMP also supported cooperation related to standardisation of data generation and exchange between as well as within the LMB countries, and provided capacity building to all parties involved. Finally, the FCMP supported the MRC in promoting the use of the established system in forest and other environmental policy making processes in the LMB.

3 The challenges

During implementation of the systems presented above, we encountered some obstacles we believe are common in developing GI based information management systems within the context of international cooperation projects. These obstacles must be seen as challenges, which have to be faced by system developers, as much as by non-technical personnel of international cooperation organization. The following section describes these challenges and gives examples on how these challenges can be met.

3.1 Challenge 1: Overcome a general reluctance of sharing of knowledge and data

In South East Asia, there is in general considerable secretiveness and reluctance to share or publish. Sharing of information and knowledge conflicts strongly with a tendency to keep secrets in order to maintain a competitive advantage (NG, 2000). Furthermore, protection of intellectual property (enacted regulations and their enforcement) is often weak in South East Asian countries, and there are no institutional provisions for the exchange or selling of data. This constitutes also strong motivations at public sector institutions to try to 'corporatise' information and knowledge and keep them under wrap, instead of managing them transparently. Under the cover of corporate secrecy, individuals are tempted to hoard not only for their institutions, but also for themselves. Hence, ownership of information and knowledge is compartmentalised, and individuals tend to keep secrets. Experience shows that this is an obstacle especially with high quality GIS data and other information that can be readily used.

Experiences in East Kalimantan showed that the different land management agencies dealing with map or GIS data were not only reluctant to share data because of ownership and financial aspects, but also because the data once combined or overlaid with additional data and information were revealing that spatial planning was based on false, incomplete or conflicting information. For example it became evident that estate crop or forest plantations licences were issued in natural forest areas, even though prohibited by law. However, vested interests and the common 'not-loosing-face'-mentality encouraged actors to maintain a status where exact boundaries of different land use systems were not known.

Reluctance to share data is even more challenging at the regional level between different countries, as evidenced during the implementation of the forest cover monitoring system in the Lower Mekong Basin. The issue may be addressed through formal data sharing agreements, the establishment of which may however take years to complete.

3.2 Challenge 2: Promote data quality

If GIS data are to be used for the implementation of sound natural resource management, these data need to be well documented, up-to-date and accurate. However, we often found that the most basic documentation of data is missing and quality assurance systems by data-producing agencies are non-existent or not implemented. This appears to be due to a number of reasons:

- the general reluctance of sharing data (see above),
- vested interests in natural resource management,
- lack of skills in IT and GIS,
- missing or badly allocated budget for quality assurance, missing consciousness on the importance of data quality.

In the case of decision support systems (such as the fire information systems in South Sumatra and East Kalimantan or the forest cover monitoring system in the Lower Mekong Basin) as well as management information systems, faulty data may cause system malfunctioning and faulty outputs. Often they are blamed on the system while the real source of the problem are the data – e.g. erroneous land cover information in south Sumatra, where e.g. an already degraded forest is still mapped as pristine leading to a low fire risk estimation for these areas. However, since the system produces aggregated information products, errors in the input data are not visible to the end users who then may wonder why an area with an assumed low fire risk suffers from frequent fires. Such faults are then readily blamed on the system while in reality they manifest errors in the underlying data.

Any international cooperation project that works with GIS data is therefore well advised to put the promotion of data quality high on its agenda. In the case of the forest cover monitoring system in the Lower Mekong Basin for instance, commonly agreed data quality guidelines and related metadata standards were established to address this issue.

3.3 Challenge 3: Don't leave the system to the technicians

Responsibility for system operation is often entirely transferred to technical staff, with the higher management level not being able to properly oversee and direct their work or assess their performance. Technical operators often do not understand the requirements for decision-making, and tend to operate the system as they see fit or even consider it as their personal domain, which negatively affects the overall usefulness of the system.

In the case of a management information system, good knowledge of methods in the field and a thorough testing of new procedures implemented by the system are required. Therefore, communication between the mid-management personnel that usually oversees software development projects and the people who actually have to produce data in the field or work with the data needs to be ensured. For the management information system developed for the forestry department on peninsular Malaysia, a comprehensive field testing of the management rules and guidelines that were developed in parallel by the Sustainable Forest Management and Conservation Project was used to test the GIS support for every planning step. During this process, valuable insight of the software implementation was gained and extensive training for ground staff was conducted.

3.4 Challenge 4: Make systems work within a weak general communications and technical infrastructure

In many developing countries, the communications and general technical infrastructure is weak (see e.g. SCIADAS, 2003). This means that one may have to cope with slow internet access (and frequent disconnections), power cuts, and limited possibilities to maintain complex systems that rely on expensive hardware (e.g. large format printers). As implementation of GIS systems proceeds from national to provincial or district level, these issues tend to become more challenging. Thus, system developers who work at national level may count on a rather well functioning GIS lab with good internet connections, while at the lower levels the intended system users do not have daily access to computers, GIS workstations or reliable internet connections. In the case of Indonesia and Malaysia respectively, implementation decisions take place on lower levels (state level in Malaysia and district level in Indonesia) where there is an extremely heterogeneous "GIS-readiness" in terms of technical infrastructure.

3.5 Challenge 5: Cope with a lack of transparency and vested interests in natural resource management

Forest and natural resource management in South East Asia is notorious for a lack of transparency and participation on the one hand and strong interests from powerful players on the other hand that hinder sound management (for Indonesia, see e.g. BARBER and SCHWEITHELM, 2001). In Malaysia, one of the main obstacles on the road to the acceptance of the Forestry Information System lay in the increased transparency and accountability provided by the system, which at times conflicted with other interests within the administration or at higher political levels. Thus, compartment boundary surveying turned out to be a rather delicate issue, because a sound surveying process allows for an exact spatial accounting that may conflict with the interest of powerful players in the forest industry and at state level. With the advent of a GIS system, which enforces the allocation of non-overlapping licenses to the entire and permanent sub-compartments, accountability and transparency are increased. In the forest cover monitoring system in the Lower Mekong Basin, combining information on forest concession, protected area boundaries and logging or forest conversion operations identified by means of remote sensing, met considerable resistance, as it was prone to detect easily boundary violations. Obviously, such systems will face certain opposition by stakeholders who see their profits at risk.

3.6 Challenge 6: Address need for institutional and human resource development

In order to effectively meet the challenges that the introduction or expansion of GIS systems pose for organizations, a strategy for institutional development is called for. One of the main lessons learned by both the IFFM project in implementing the Fire Information System for East Kalimantan and the forest cover monitoring system in the Lower Mekong Basin, was that transfer of technical knowledge through training for individuals is not sufficient to achieve sustainability in project goals. These activities have to be embedded in a broader process that includes development of roles and responsibilities for system operation and maintenance, development of budget plans (e.g. to handle all costs of maintaining a running system, which are often not assessed at the onset of development); and the institutionalization of knowledge acquired by individuals during the development phase to transfer knowledge from individuals to the organization. That means to increase capacity of these skills and knowledge in an organization one has to assess also the organizational structure (management and decision-making processes, working mechanisms) and the regulatory frameworks that might hamper implementation.

3.7 Challenge 7: Increase ability of organizations to maintain and expand systems after the international cooperation project ends

The moment when an international cooperation project ends and the experts leave is critical for the project success. Organizations need to be prepared to handle the disappearance of funds as well as of expert knowledge from this moment onwards. Therefore, a plan for maintenance and expansion of a GIS system needs to be developed to secure both knowledge and funding for the system to live on. The IFFM project has in its last phase focused on the capacity building on the system and institutional level to make efforts of technical knowledge transfer and application sustainable. This enabled the partner institutions to carry on with further use and development of the Fire Information System not at least by having achieved an institutional structure for fire management, but also creating an enabling environment for fire management that can function within the Indonesian socio-economical and political-administrative framework.

3.8 Challenge 8: Make project results available to stakeholders

Availability of information products has two components: (1) the product is physically available, and (2) it can be used by the intended user – i.e. the product must be intelligible, in a format ready for use (e.g. on paper, as a jpg image, as GIS data or as information on a web page according to the user's needs). From what has been said previously, it is clear that in a country like Indonesia, with a weak communications structure, limited human and institutional resources and against a background of a general reluctance to share and publish data, even getting the information product to stakeholders may be quite a challenge. A study to evaluate the Fire Threat Analysis application in South Sumatra (KOLLAR 2006) revealed that a large number of the intended recipients of fire risk maps stated that they did not receive the maps although they were sent as paper prints to their respective agencies.

The second point – usability-refers to the distribution format of project outcomes – e.g. maps or GIS data. In the case of a rather complex decision support system like Fire Threat Analysis the main challenge is to keep the output simple and understandable and with a clear application in mind. There may be a high reluctance to manage information and knowledge with no clear application; however, envisaging clear applications may be particularly challenging in complex situations characterised by high levels of uncertainty, such as in natural resource management (BUCHHOLZ et al., 2005).

4 Synopsis and conclusions

From the experiences outlined above, we can draw the major conclusion that setting up and implementing an efficient and sustainable GIS system is not only a technical or scientific issue, but is putting a high demand on the organisational as well as communication culture of the involved organizations – both development organization and partner.

When designing a GIS system, aspects of institutional and organizational development have to be considered from the very beginning, meaning that a comprehensive capacity building effort needs to be included in the project plan. When looking at the challenges described above, it is clear that 'capacity building' is far more than training of individuals but has to take place on three levels to be effective and sustainable (see e.g. ANONYMOUS, 2003):

- the systems level, e.g. the regulatory framework, policies and frame conditions that support or hamper the achievement of certain policy objectives
- the institutional or entity level, i.e. the structure of organizations, the decision-making processes within organizations, procedures and working mechanisms, management instruments, the relationships and networks between organizations
- the individual level, i.e. individual skills and qualifications, knowledge, attitudes, work ethics and motivations of the people working in organizations.

In order to overcome the challenges described above, we identified a number of critical non-technical elements for system development that we feel need to be considered when planning and implementing GI projects in natural resource management:

1. While the actual system development team should be kept small and assigned with clear tasks, relevant stakeholders including management and the policy making levels need to be integrated into the system development and implementation process at an early stage to gain broad acceptance and participation.
2. System development needs to be a joint effort between development organization and the partner. This requires a concise GIS project definition as a result of an initial assessment and analysis phase to ensure agreement as to where the journey shall go. Bringing local consultants or partners into the development team may foster system sustainability.
3. Clearly define the key benefits of the system with the stakeholders. The product output of the system should be clear, and the products need to fit the use of the stakeholders. This includes also facilitation and institutionalisation of distribution pathways.

4. Institutional strengthening and capacity building need to form elements of information systems development right from the beginning to enable the independent use and further development of the established system.
5. The complexity of an information and management system needs to be adapted to the resources available to ensure the long-term sustainability of the system. This requires a thorough assessment of the existing capacities prior to defining the project design.
6. Care needs to be taken to ensure availability and quality of the data forming the core of the system, e.g. to develop mechanisms for data quality management, data exchange and data documentation.
7. Enough resources (financial as well as human) have to be allocated within the project team to plan and supervise the system development. This requirement is valid for both the development organization and the partner.
8. The system needs to be developed in short, clearly defined stages over a sufficiently long period of time to allow for learning and institutional development processes within the involved organizations. This allows an early, stepwise implementation of the system, effective monitoring and evaluation of the entire process and increased flexibility in development.

These points may be further elaborated to form an initial framework for strategic planning and evaluation of GIS components within the context of development cooperation projects.

References

ANONYMOUS, 2003, InWent Capacity Building Concept. Available online under: http://www.inwent.org/ publikationen/material/index.de.shtml

BARBER, C.V. and SCHWEITHELM, J., 2000, Trial by Fire. Forest Fire and Forestry Policy in Indonesia's Era of Crisis and Reform. World Resources Institute, Washington.

BOWEN, M.R., BOMPARD, J.M., ANDERSON, I.P., GUIZOL, P. and GOUYON, A., 2001, Anthropogenic Fires in Indonesia: A view from Sumatra. In EATON, P. and RADOJEVIC, M. (Eds.), Forest fires and Regional Haze in South East Asia, Nova Science, New York.

BUCHHOLZ, G., CAMPHAUSEN, A. and FELDKÖTTER, C., 2005, Information and Knowledge Management Systems within the Context of Natural Ressource Managenet in South East Asia. Experiences from 25 years of German Technical Cooperation. GTZ, Eschborn. 36 pp. http://www2.gtz.de/publikationen/isissearch /publikationen.

CHOKKALINGAM, U., KURNIAWAN, I. and RUCHIYAT, Y., 2005, Fire, livelihoods and environmental degradation in the Middle Mahakam Peatlands. Ecology and Society, 10, Article 26. http://www.ecologyandsociety.org/vol10/iss1/art26/.

HOFFMANN, A.A., 2003, A Fire Information System for East Kalimantan, Indonesia, International Forest Fire News No. 23 – December 2000, pp. 29–32.

KOLLAR, I., 2006, Quality assessment of a vegetation Fire Threat Analyss System developed within the framework of a development cooperation project in South Sumatra, Indonesia. Diploma theses, Ruprecht-Karls-Universität Heidelberg, 176 pp.

NG, F.S.P., 2001, Biodiversity and Asian Attitudes on Knowledge. In KHENG, C.T. and LEE, E.Y.E. (Eds.), Biodiversity 2000. Proceedings of Intern. Conference on Prudent Biodiversity Management and Sustainable Development Sarawak Biodiversity Centre, pp. 169–176.

RÜCKER, G. and BUCHHOLZ, G., 2003, Ein GI-System zur nachhaltigen Bewirtschaftung tropischer Regenwälder in West-Malaysia. In STROBL, J., BLASCHKE, T. and GRIESEBNER, G. (Eds.), Angewandte Geographische Informationsverarbeitung XV. Beiträge zum AGIT Symposium 2003. Wichmann, Heidelberg, pp. 397–403.

SCIADAS, G. (Ed), 2003, Monitoring the Digital Divide … and beyond. NRC Press, Montreal, 182 pp. http://www.orbicom.uqam.ca.

SIEGERT, F., RÜCKER, G., HOFFMANN, A. and HINRICHS, A., 2001, Increased damage from fires in logged forests during droughts caused by El Nino. Nature, 414, pp. 437–440.

Middle East Water Challenge – Supporting Negotiation Processes by Land Use Change Modelling: Estimation of Land Use Change Probability Patterns, Considering Alternative Water Supply Policies

Wolfgang LOIBL and Ernst GEBETSROITHER

Summary

The chapter discusses research carried out within the EU funded project. 'A Future of the Dead Sea Basin' (*INCO-MED contract no ICA3-CT-2002-10019)* as preparation for applying a land use change model – combined with a system dynamics model (GEBETSROITHER, this volume). The approach supports scenarios on future land use considering the entire Dead Sea basin, regardless of political and administrative borders. The chapter provides some insights in the current political framework with its territorial and water claims and leads then to methodological issues to conduct spatially explicit land use change probability patterns. These patterns are estimated via regression models based on current land use patterns related to 'natural' landscape characteristics. The tasks consider alternative land use development as effect of certain political interventions: e.g. the 'Red-Dead-Peace Conduit' delivering water from the Red Sea to the Dead Sea basin and the Dead Sea itself. To integrate these effects, 'natural' land use change probability layers are finally merged with additional layers considering change of spatial attractiveness due to political interventions triggering dwelling-, industrial-, tourism- and crop cultivation-activities.

1 Political framework: establishing territorial claims in the Middle East through water claims

The Middle East area with the neighboring countries Lebanon, Syria, Israel, Palestine and Jordan are all depending on the waters within the Jordan River catchment area. The upper Jordan River and its confluences discharge water to Lake Tiberias serving as regional water reservoir. Lake Tiberias and Yarmouk River both discharge around 500 Million cubic meters per year (MCM/yr) to the lower Jordan River. Due to diversions of the Jordan- and the Yarmouk-waters only little water reaches the Dead Sea. Discharge has dropped from 1250 MCM/yr freshwater to 200 MCM/yr of sewage (ORTHOFER et al., 2001). The Dead Sea inflow deficit reaches 1000 MCM/yr and an exaggerated 'outflow' due to forced evaporation by mineral extraction activities reaches 250 MCM/yr. Thus the Dead Sea level and the groundwater table declines of the past 50 years – currently estimated at about 1 meter per year.

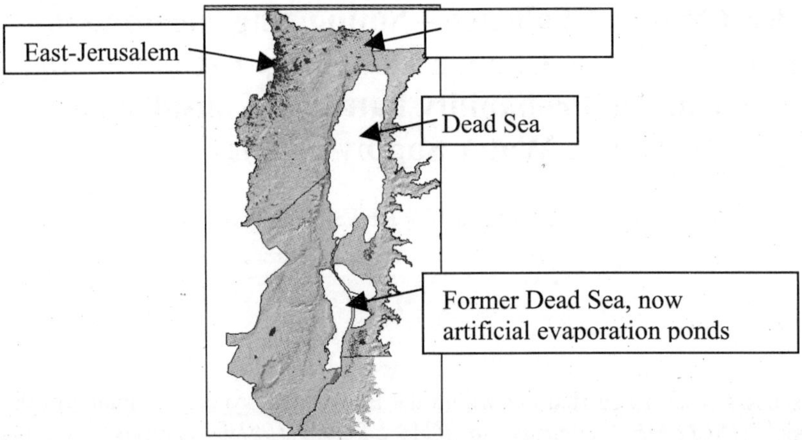

Fig. 1: Study area: Settlements (red) and crop cultivation allocation (green), Palestine territory: north, Israel territory: south, Jordan territory: east

The relation between water and territorial claims is evident and water scarcity represents a key pressure component in the region since long. Thus, we shortly illustrate the historical political framework: Soon after the end of World War II, in 1948, the British protectorate Palestine has declared itself as the independent state of Israel. Within a year, some 100 000 Palestinians have been displaced from their former homes and Jewish refugees from around the world where invited to immigrate. As part of the Zionist ideology of developing the Palestinian desert into an 'agricultural heaven', Israel wants to provide new settlers with an 'abundance' of water to allow for a prosperous life. Israel resumed water planning soon after the independence declaration in 1948. US ambassador Eric Johnston mediated in 1953 a development plan for the Jordan River system, to reduce the conflict potential by committing water quotas for all riparian states. As the Johnston Plan failed, both Jordan and Israel carried out two major successful projects: The Israeli National Water Carrier diverted initially 320 MCM/yr water from Jordan River to the entire Israel and Jordan's East Ghor Canal which delivers around 500 MCM/yr water from Yarmouk River to supply the Jordanian Jordan Valley, the Dead Sea area and Amman (NAFF and MATSON, 1984).

The Palestinian population in the Westbank suffers since long from lacking water for domestic use and irrigation. The Israeli water authority Mekorot delivers water to the entire community of Jerusalem and also to neighboring Palestinian highland habitations. For the Jordan Valley and Dead Sea region additional 270 MCM/yr ground water of sufficient quality is abstracted (ORTHOFER et al., 2001). Israeli settlements and plantations are provided with enough water through wells drilled to a depth of some 100 meters, whereas Palestinian dwellers are not allowed to drill deep wells. Thus, besides the restricted and costly Mekorot supply, they are dependent on informal water sources like local springs and wells, abstracting shallow (lower quality) groundwater and surface water harvested by cisterns.

2 Regional policy, spatial planning and land use management

Land use management has to be considered in a local context and embedded in the legal framework and informal rules of the respective countries. In western countries the major mission of regional policy is, to mitigate regional disparities – e.g. between prosperous central regions and deprived remote areas – or to strengthen countries' growth poles to support the economic competition. Planning laws and development plans control land use to avoid conflicts between adjacent land uses or to regulate land use densities. As land in densely occupied western countries is a scarce commodity, land use management refers to preservation of open space and nature conservation by constraining development of built-up land through zoning regulations. Traffic infrastructure is still decided and built by national authorities, but land use management is mostly left to the private sector, controlled by law. Water is rarely a matter of discussion as there is usually enough water available.

In Middle East countries regional policy is, at least in remote areas, focused on territorial policy to secure national claims on land and water. Because water is a scarce commodity in arid areas, water supply is a national issue. In those remote regions most area is 'un-used' open space and land use management concentrates on establishment of dwellings and cultivation of fields and plantations. Nature conservation is still of less importance. Because land is no limiting factor, planning regulations are somehow weak and land use management is more dependent on – nationally controlled – water availability.

Land use is managed in the Jordan Valley/Dead Sea area differently by the 3 states:

Israel is developing the area from a strategic viewpoint establishing new 'settlements' as colonies in Palestinian territory between Palestine dwellings to claim Israeli territorial rights. Agricultural development is performed systematically and efficiently through large plantations. Land use management at local level is indirectly maintained by national authorities, left to reliable Israeli developer-companies to establish 'urban settlements' or left to Kibbutz-settler-communities or agro-companies for export-oriented crop cultivation

Palestinian land use management is dependent on Israeli concessions regarding accessibility and water rights. Dwelling development is left to the private sector, nevertheless depending on the regulations of the Palestinian water authorities and 'silently' controlled by Israeli authorities. Farming remains on a small scale, it is privately organized and highly dependent on informal water supply from wells, springs and surface water, harvested through cisterns.

In Jordan land use management along the eastern Jordan Valley and Dead Sea is usually privately handled, but more or less depending on formal water distribution provided though network infrastructure. Land use change regarding construction of houses or crop cultivation plantations refer to private activities not heavily regulated by planning guidelines but by water distribution.

3 Land use change probability modelling

Following this hydro-political framework, land use change scenarios were developed within the study, and spatial allocations of new land uses were estimated.

Land use change modelling has been started 50 ago with certain progress in terms of spatial accuracy and resolution during the last 15 years (c.f. ALONSO, 1964; TURNER et al., 1995; LAMBIN and GEIST, 2001; BENENSON and TORRENS, 2004; LOIBL et al., 2007). In general 2 different approaches are used:

- estimation of land use class change proportions for entire regions or sub-.regions or
- estimation of cell-related land use change within a raster-layer proving results for very local patches within a continuous landscape

To derive spatially explicit results the latter approach was conducted through agent-based modelling (ABM). ABM is a rather new approach to simulate future land-use change as effect of spatial activities of individual agents. ABM addresses single agent's actions through iterative steps leading to steady land-use changes over time. Landscape properties are integrated into those models as raster-cell-layers. Details can be found in GEBETSROITHER and LOIBL (2006; this volume).

Land use change and water demand were related to human activities triggered by landscape characteristics. Areas selected more often as action destination (for settling, for crop cultivation) than others show a higher land use change related activity probability. As explained earlier, simulation of future land use change has to refer to certain political interventions, leading to infrastructure allocation which control development in certain areas. Because allocation made by interventions did not fully consider land use suitability due to strategic or other reasons and allocation of future intervention is somehow predefined through (hydro-)political scenarios, land use change is not that closely related to landscape characteristics. Land use change modelling has to integrate the spatial effects of interventions over-ruling 'natural' land use change-related activity probability. Thus the entire land use change probability has to be derived from the spatial influences assumed as land use change attractiveness and allocation of interventions:

- topology (distance and relation between spatial elements: roads, shore, settlements, etc.),
- topography (elevation and terrain characteristics, infrastructure (road, pipeline network),
- availability and location of natural resources – overall water sources – and
- allocation of spatial interventions (e.g. infrastructure development).

To estimate land use change probability several single attractiveness layers were explored regarding their ability to explain land use change. This has been carried out via multiple regression models, analysing relevant landscape property criteria and quantifying their contribution to the overall influence on land use change, explaining allocation and intensity of – e.g. population settling and crop cultivation. Several raster layers were input for generating attractiveness layers: elevation, slope angle, road and water distribution network and land-use classes (habitations, industrial sites, hotel sites, crop cultivation types). A set of several land use attractiveness layers were generated and tested by regression analysis whether they explain the land use patterns. Most of them are distance layers: to habitation, crop cultivation area, industrial site, hotel site, road- and water network and the Dead Sea shore. Additional layers contain water supply information: water supply potential reachable within 5 km, 'supplied area' around each water source dispersing the locally available water quantity into the surrounding area considering water demand of crop cultivation and domestic requirements per area. The calculation of local attractiveness was conducted using the land use related raster layers and raster GIS functions (find distance, focal-mean and focal-sum). Figure 2 shows some land use change attractiveness layers.

To quantify the influence of the different attractiveness layers on land use change, regression analyses were performed with population numbers and crop cultivation area (per 1×1 km sample cell) as dependent variables and spatial attractiveness data as independent variables. The data sets were extracted from the raster-layers taking values of the cells from all layers with population numbers and crop cultivation area and of a certain number of random cells without population related activity.

The right-most map in figure 2 shows the sample-cell selection: habitation-centres crop cultivation patches and additional random cells without population related activities, to estimate regression functions. The resulting surfaces were calculated by applying the regression functions: each regression coefficient was multiplied by the cell values of the related raster-layer representing the respective independent variable which (co-)explains population number or crop cultivation area of the certain cell as dependent variables. Finally the resulting raster-layers reflecting the regression equation elements were summarized to achieve the regression model results. The absolute value results were normalized to meet a 0–1 range, aiming in spatial probability activity surfaces for settling and for crop cultivation as activity allocation pattern for the ABM.

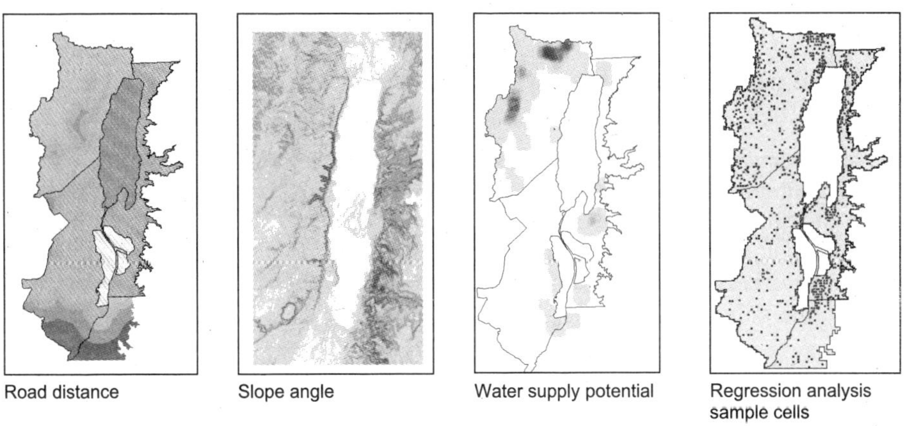

Fig. 2: Sample attractiveness layers and sample cells to conduct the land use change probability analysis

4 Land use change probability surfaces considering general land use change suitability and effects of political interventions

Earlier we have discussed political interventions over-ruling natural land use change probability patterns. The interventions integrated here are defined by certain Dead Sea future scenarios. Some of these scenarios refer to the construction of a water-pipeline from the Red Sea to the Dead Sea – the 'Red-Dead Peace Conduit' – along the Jordanian border that should deliver additional 850 MCM drinking water/yr. This intervention will have great effect on the southern Dead Sea basin and could mitigate water scarcity for all riparian states; despite of this, there are certain disadvantages related to the hydro-chemistry of the Dead Sea water. To integrate spatial effects of political interventions into the 'natural' probability surfaces, additional trend surfaces have been derived considering additional water potential, slope angle and distance to the future 'Red-Dead Peace Conduit', where the normalized results are added to the original probability surfaces re-weighting them to carry out new activity probability patterns for different destination areas.

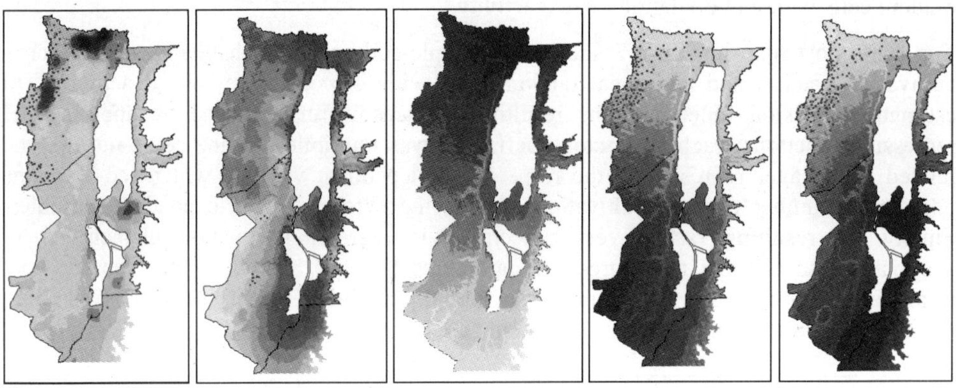

Fig. 3: Final activity probability surfaces leading to land use change

Figure 3 shows land use change probability surfaces: settling, crop cultivation, further re-weighted probability patterns for Palestinian activities, for Israeli activities and for integration of regional effects of the "Red-Dead Peace Conduit" in the southern Dead Sea basin. The darker the colour the higher is the probability for future action. Those probabilities have been finally applied within the ABM described in GEBETSROITHER and LOIBL (2007, this volume).

5 Conclusions

Modelling land use change probability patterns has two advantages:

(1) it allows a fast and simple visualisation of general allocation trends of future local land use – related activities based on and quantified through landscape properties, even if no further detailed simulation is conducted,

(2) it delivers a traceable spatial perception-background to control actions of agents within ABMs for spatially explicit simulation of activities leading to land use change, examined through models that consider the certain dependencies of human activities on landscape properties.

By including layers that reflect allocations of singular political interventions like establishing large infrastructure projects, the local effects of regional development strategies can be additionally integrated into the perception pattern for the agents that attract or repulse human activities.

This approach, providing a wide range of activity probability layers which lead to land use change, allows flexibility in simulating spatial effects of various development scenarios with high local certainty.

References

ALONSO, W., 1964, Location and Land Use. Cambridge, Mass., Harvard University Press.

BENENSON, I. and TORRENS, P.M., 2004, Geosimulation: Automata-based Modeling of Urban Phenomena, Wiley, Chichester.

GEBETSROITHER, E. and LOIBL, W., 2007, GIS based water resource management of the Dead Sea region – integrating GIS, System Dynamics and Agent Based Modelling. In ZEIL, P. and KIENBERGER, S. (Ed.), Geoinformation for Development. Wichmann Verlag, Heidelberg, pp. 26–32.

LAMBIN, E.F. and GEIST, H.J., 2001, Global land-use and cover change: What have we learned so far? Global Change Newsletter, 46, pp. 27–30.

LOIBL, W. et al. 2007, Simulation of polycentric urban growth dynamics through agents – Model concept, application, results and validation. In KOOMEN, E., BAKEMA, A., STILLWELL, J. and SCHOLTEN, H. (Eds.), Modelling Land-Use Change. Springer Netherlands, pp. 219-235.

NAFF, T. and MATSON, R.C., 1984, Water in the Middle East: Conflict or Cooperation? Westview Press. Boulder, Colorado, USA, and London, pp. 1–62.

ORTHOFER, R. et al., 2001, Developing Sustainable Water Management in the Jordan Valley. Joint Synthesis and Assessment Report of INCO-DC Project No ERBIC18CT970161. Report to European Commission, Sep 2001, ARC Seibersdorf research.

TURNER II, B.L et al., 1995, Land use and Land-Cover Change Science and Research Plan. IGBP Report No.35, Global Change Newsletter-annex, ISSN 0284-805. Geneva.

GSE Global Monitoring for Food Security (GMFS) – "Food Security Information Services for Africa"

Carsten HAUB

Summary

GMFS is a mostly satellite-based data support service for existing early warning systems. The main aim of GMFS is to provide advanced crop information derived from earth observation data, supplemented by the use of agro-meteorological models. GMFS aims not only to provide earth observation based services but also to encourage partnerships in monitoring Global Food Security and related environmental processes, by concerting efforts to bring data and information providers together, in order to assist stakeholders, nations and international organizations to better implement their policies towards sustainable development. In particular, GMFS aims at contributing to improved food security in the African continent by providing additional remote sensing based information to organizations at the national, regional and international level.

Main objectives of GMFS are (i) to improve the provision of operational and sustainable information services, (ii) to provide advanced crop information derived from Earth Observation data, (iii) to assist food aid and food security decision-makers at the local to global level, and (iv) to contribute to the transparency of the production, management and distribution of agriculture.

1 Introduction

In June 2002 the World Food Summit reaffirmed its pledge to end hunger and called for an international alliance to accelerate efforts towards that goal. In line with the Millennium Development Goals, the summit participants aimed to reduce the number of people without enough food to eat from the 800 million existing today to about 400 million by 2015.

This general goal has been adopted by the EC and EU member states in their development policies. In direct support to the objectives of these policies the GSE Food Security portfolio aims to provide sustainable operational information services, derived at least partly from earth observation data, to assist food aid and food security decision makers.

The target user community of the GSE Food Security services are decision makers in non-governmental, governmental and intergovernmental organisations who require information on agricultural production in the food-insecure countries of sub-Saharan Africa. Direct users of the services are primarily the existing agencies who already support food-security decision makers (FAO, WFP, JRC/MARS-FOOD, Regional and National Early Warning Systems, national ministries of agriculture, etc.), rather than the decision makers themselves. This strategy aims to avoid duplication of existing services, such as those provided by FAO/GIEWS and JRC/MARS-FOOD. It also allows the service network to concentrate its resources and expertise on EO-based information, rather than on the integration with socio-economic data and subsequent expert interpretation.

GMFS (Global Monitoring for Food Security) is a GMES Service Element project, part of ESA's contribution to the EU/ESA GMES Program (Global Monitoring for Environment and Security). It aims to establish an operational service for crop monitoring in support of Food Security Monitoring to serve policy makers and operational users. The GMFS project started in March 2003 as part of the Stage 1 of the ESA Earthwatch GMES services Element 'Service Consolidation Actions', which ended in December 2004. The GMFS project was continued from Stage 1 in October 2005 as part of the Stage 2 of the ESA Earthwatch GMES services Element – 'Scaling Up Consolidated GMES Services' and is scheduled to run for 3 years (up to September 2008).

The GMFS consortium has been set-up taking into account the specific experience of the involved partners and making optimal use of the networks they are operating in. Twelve partners are involved in the project: VITO (B) as Prime Contractor, ESYS consulting (UK), GIM (B), TRASYS SPACE (B), SARMAP (CH), Synoptics (NL), AVIA-GIS (B), ULG (B), ITA (IT), FMA (IT), EARS (NL) and EFTAS (D).

The consortium brings together expertise in (i) development and delivery of operational services, in particular with respect to monitoring of crop conditions; (ii) the public domain policy analysis; (iii) the early warning and crop monitoring user community and (iv) development and analysis of sector infrastructure. The consortium is supported and strategically guided by a service strategy group consisting of authorities from the policy and user sector (FAO, JRC, WFP). Additionally the GMFS consortium is guided by a User Executive Board consisting of user-federating agencies/institutes at the international (FAO, JRC, WFP) and regional (AGRHYMET, RCMRD, SADC-RRSU) level.

GMFS will provide monitoring services regarding crop state/vegetation condition at continental scale and national scale. At continental scale crop state/vegetation condition is monitored using low resolution Earth Observation (EO) data, while at national scale products will be generated using medium and high resolution data, field work and modelled products (AMM). National scale activities will cover Senegal and Malawi in continuation of activities in 2003–2004 and Ethiopia, Sudan and Zimbabwe as 'new' GMFS countries.

2 GMFS services portfolio

GMFS services are tailored to meet the information requirements of core users. The initial portfolio, as defined in detail by the end-users, consists actually of two services:

- Regional/continental monitoring and forecasts, aiming at monitoring continuously, primarily at continental and regional scale (1:2 000 000–6 000 000) – usually on a 10-days (i.e. average 10-days images) basis – the actual rainfall, and vegetation changes by analysing long-term EO time-series.
- National/sub-national monitoring and forecasts, mainly aiming at delivering detailed information on actual acreage during the agricultural season for major crop types – at planting date, during the crop growth, and at harvest time, provided at scales ranging from 1:50 000 to 1:500 000.

GMFS will furthermore support Food Security Monitoring activities at regional and national level through:

- Making available regional-level crop state indicators on a regular basis allowing to assess the progression of the current growing season.
- Analysing satellite data aiming at generating acreage estimates on particular areas based on the priorities set by national authorities and derived from satellite data (high resolution, 15–30 m)
- Analysing and processing satellite data to generate yield estimates for major crops based on different methodologies, on a regular basis (10-daily) throughout the growing season;
- Implementing agro-meteorological models, based on ground observations, meteorologically data and ancillary data to generate yield estimates for the main crops at the end of the season
- Providing the produced products in a timely manner to stakeholders and training the relevant national counterparts staff members in correct usage and interpretation of the products
- Providing improved/alternative options should new techniques become available for routine application or current remote sensing platforms fail to produce data
- Support the national counterparts in carrying out fieldwork surveys in relation to the split sampling approach where one part serves as a training set to feed satellite data driven forecast models
- Providing advise on standardized reporting with regard to food security in specific countries and on the strategic use of Remote sensing in the overall food security monitoring process
- Supporting the national stakeholders in their training activities in particular in relation to the validation, interpretation, processing and utilization of (remotely sensed based) information as per the adopted integrated methodology

Additional important parts of the GMFS Service are the Ground Products, i.e. the collected field data used for validation purposes and as input for the agrometeorological model. Capacity building is provided on a technical level via (user-tailored) training sessions on the service as well as on the production techniques, and on institutional and political level by raising the awareness about the benefits of the service in the decision- and on policy-making process to reach a high-level institutional commitment.

3 GMFS methodology

Regional and continental monitoring and forecasts

1. Yield maps of Sub-Saharan Africa are produced using low-resolution satellite imaging.
2. General production figures are calculated using the yield maps, general agricultural zones (also derived from low-resolution satellite imaging) and agro-meteorological models (AMM). The models determine daily crop growth based on rainfall, sunshine duration, sowing date and the amount of water in the soil.
3. This data is compared with yield and production data from past assessments to evaluate prevailing trends.

National and sub-national monitoring and forecasts

1. Detailed crop acreage maps are derived from multiple high-resolution images captured over time.
2. Fieldwork missions are conducted to collect ground data for the validation of the acreage maps and as input into the agro-meteorological models.
3. Agro-meteorological models are again employed to calculate daily crop growth. This generates the yield per crop.
4. The crop acreage and yield per crop are used to calculate detailed production figures.
5. These high-resolution results are used to fine tune and enrich the low-resolution surveys.

The GMFS methodology optimally combines low resolution and high resolution satellite data. Low and medium resolution data are used for continuous monitoring on a global and regional scale. High resolution data are used to zoom into problem areas and also to calibrate the low resolution data.

4 Users

The initial focus of GMFS is Sub-Saharan Africa. GMFS feeds its information into existing regional, national and international early warning systems that monitor agriculture for food security. These direct beneficiaries, such as the Centre de Suivi Ecologique and the Southern African Development Community, advise decision-makers at development and food aid agencies, and policy-making and research institutions (indirect beneficiaries).

In the framework of the current project, GMFS is working closely with the following organisations:

- Southern African Development Community SADC / Regional Remote Sensing Unit RRSU, Gaborone, Botswana
- Centre de Suivi Ecologique – CSE, Dakar, Senegal
- Malawi, Ministry of Agriculture, Irrigation and Food Security
- EC-JRC MARS-FOOD project
- World Food Programme (WFP)
- Food and Agriculture Organisation (FAO)
- Direction de l'Analyse, de la Prévision et des Statistiques (DAPS), Dakar, Senegal
- CILSS-Agrhymet Regional Centre
- Regional Centre for Mapping of Resources for Development (RCMRD), Nairobi, Kenya

5 Examples

5.1 Low-resolution indicators

Dry Matter Productivity or DMP reflects the increase of dry matter biomass on a daily basis (kgDM/ha/day). When crops are healthy and water and nutrients are not limiting, dry matter production is proportional with the amount of light intercepted by a crop canopy

(MONTHEITH, 1972). Incoming shortwave solar radiation on average comprises 48% of Photosynthetically Active Radiation (PAR). A fraction of this PAR is absorbed by the green vegetation (fAPAR) and converted into biomass (radiation use efficiency) taking into account losses related to the transport of photosynthetates, maintenance of the standing phytomass. The efficiency of this conversion is temperature-dependent. Using the classical Monteith-model which combines the remote sensing imagery (fAPAR, the fAPAR values can be estimated from satellite imagery, either using the Gobron (see Gobron et al.) method or they can be converted from NDVI) with meteorological data in order to obtain estimates of the productivity of the terrestrial vegetation can be made. The dry matter productivity images are calculated based upon the integration of daily meteo-data and MERIS FR/RR images.

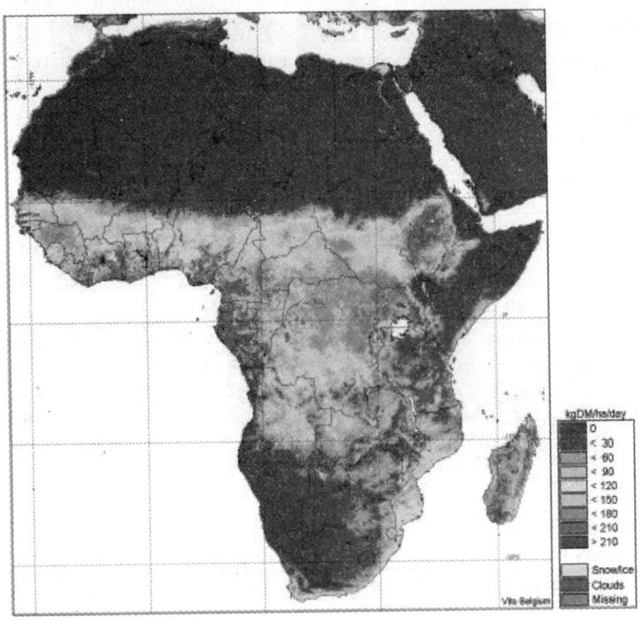

Fig. 1: Dry Matter Productivity – Netto Primary Productivity (measure of standing biomass) on 10 daily basis (VITO © 2006)

The *Vegetation Productivity Indicator* is used to assess the overall vegetation condition and is a categorical type of difference vegetation index, referenced against the NDVI percentiles of the historical year. The VPI is classified in 5 main groups. The VPI method is a statistical distribution of the NDVI for each 10-day period of the year by applying techniques commonly used in hydrology for the prediction of extreme events. The VPI can be calculated based on DMP (Dry Matter Productivity) or NDVI values. The VPI can be stratified to remove effects of the main vegetation types in order to assess vegetation status. The stratification can be done based on main vegetation types based on best available landcover. Currently the VPI is produced based on SPOT-VGT NDVI and DMP values. From the VPI images a statistical database is calculated based on the GLC2000 landcover database for Africa and UNEP administrative boundaries database. The VPI method was originally developed by SANNIER et al. (1998-2) based on NOAA AVHRR data for a study area in Zambia, and later on implemented by EERENS et al. (2000) for Europe (for MARS-STAT) and Africa (MARS/FOOD-GMFS) based on SPOT-VEGETATION data. Currently, the whole of Africa is processed based on SPOT-VEGETATION data, individual countries/areas are extracted from the Africa wide products. The purpose is to identify drought affected areas. The VPI is used to qualitatively identify areas with potentially low agricultural productivity as compared to what can be expected based on the historical range. Percentages indicate the probability of getting a lower NDVI value based on historical analysis of the NDVI values. Thus, a probability of 50% indicates that there is a 50% chance of getting a lower value (and thus 50% chance of getting a higher value) compared to the historical value range, indicating a fairly normal/average situation. A 20% probability level indicates that there is a 20% chance of getting a lower value (and thus 80% chance of getting a higher value) compared to the historical value range, indicating a worse situation compared to the previous years for the given date.

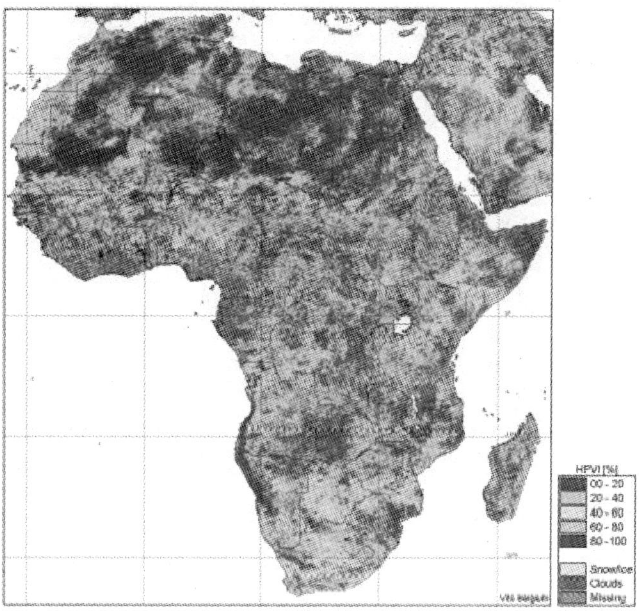

Fig. 2: Vegetation Productivity Index – Difference of vegetation growth with historical year (SANNIER et al. 1998-1). Also supported on 10-day basis: Relative diff, Relative range (VCI) (VITO © 2006).

5.2 Medium-resolution classification

The extent of cultivation products intends to give a qualitative indication of agricultural/planted areas for a given growing season. The extent of cultivation products is a repetitive product which is generated every growing season on selected areas. It is based upon medium resolution image data (pixel size 100–300 m), in specific SAR (Synthetic Aperture Radar) data in combination with multi-temporal optical data. The product is based on the multi-temporal characteristics of vegetation, and in specific the changes of agricultural land during the growing season (field clearance, sowing, senescence, harvesting, etc.). This requires repetitive acquisitions of satellite data during the growing season, typically once every 2 weeks. The extent of cultivation product covers larger areas, potentially whole countries, depending upon its size. The main difference with the cultivated area product is the scale (less detail due to lower resolutions base data) and the additional use of multi-temporal optical data.

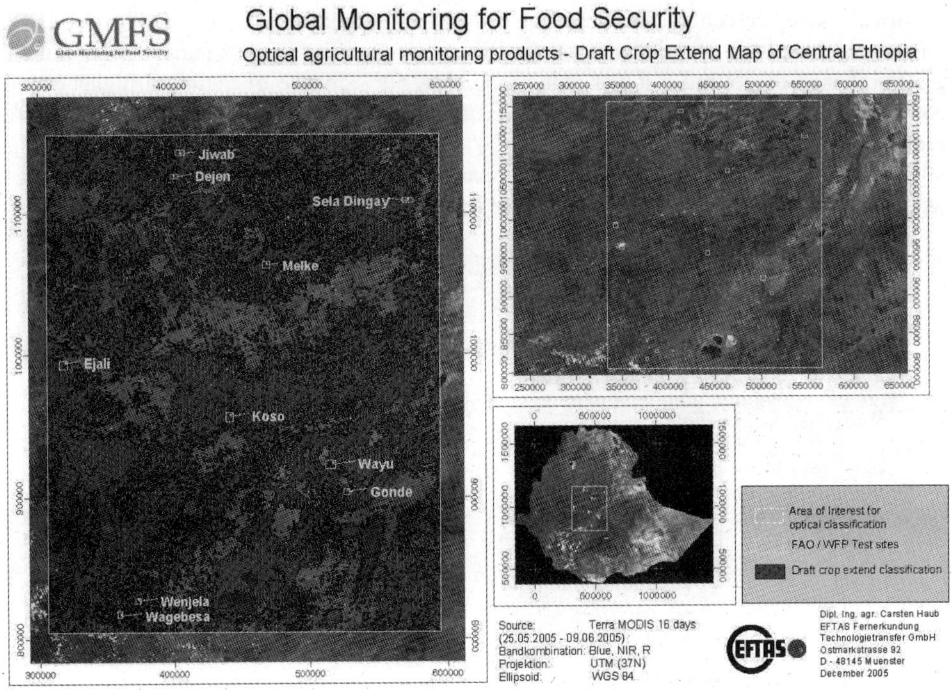

Fig. 3: Classification result: Crop extend map based on MODIS 16 days composite, SRTM and topographical maps 1:200 000 (EFTAS © 2006)

5.3 High-resolution classification

The Crop Emergence Product gives an indication on the date of emergence of the crops in a given area. It is based upon multi-temporal acquisitions of SAR data (or optical), starting well in advance of the growing season. Reported emergence dates refer to the date of satellite data acquisition when a change in vegetation activity was noted. Actual emergence date fall in between the date when vegetation change was noted on the imagery and the previous acquisition. Data is acquired typically every 1–2 weeks at the start of the growing season. The emergence date product is a repetitive product which is generated every growing season on selected areas/countries. It is based upon medium resolution image data (pixel size 100–300 m) or high resolution image data (< 30 m) for localized areas, in specific SAR (Synthetic Aperture Radar) data. The product is based upon the multi-temporal characteristics of vegetation during pre-planting and planting time (onset of rains, field clearance, sowing, senescence). Important to note is that the crop emergence date product is generated half way through the season using the same processing chain as for the extent of cultivation product.

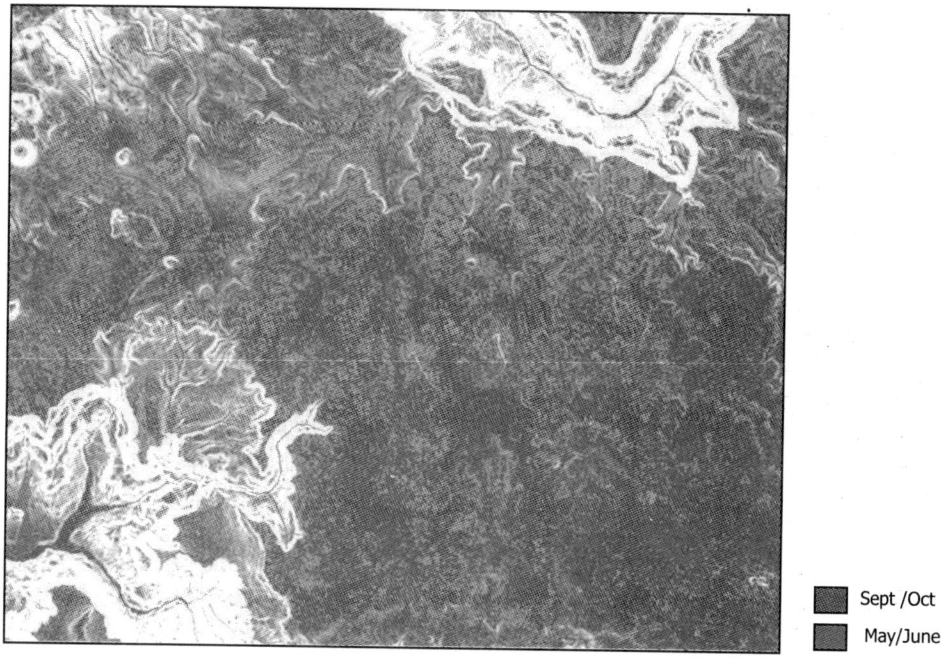

Fig. 4: Classification result: Crop emergence map based on multi-temporal SAR (ASAR AP) (SARMAP © 2006)

References

BOOGAARD H., EERENS, H., KEMPENEERS, P., PICCARD, I., SUPIT, I., VAN DIEPEN, C. and VERHEIJEN, Y., 2002, Description of the MARS Crop Yield Forecasting System (MCYFS). METAMP-report 1/3, Alterra and VITO, JRC-contract 19226-2002-02-F1FED ISP NL.

EERENS H., WOUTERS, K., BUFFET, D., OGER, R., DEHEM, D. and TYCHON, B., 2000, Use of 1km² resolution imagery in the Belgian Crop Growth Monitoring System (B-CGMS), Proc. Conf. VEGETATION-2000, 3–6 April, 2000, Belgirate, Italy, pp. 361–366.

MONTEITH J.L., 1972, Solar radiation and productivity in tropical ecosystems. J. Applied Ecology, 19, pp. 746–766.

MYNENI R. and WILLIAMS, D., 1994, On the relationship between fAPAR and NDVI. Remote Sensing of Environment, 19, pp. 200–211.

SANNIER C., TAYLOR, J., DU PLESSIS, W. and CAMPBELL, K., 1998-1, Real-time vegetation monitoring with NOAA-AVHRR in Southern Africa for wildlife management and food security assessment. International Journal of Remote Sensing, 19(4).

SANNIER C., TAYLOR, J., DU PLESSIS, W. and CAMPBELL, K., 1998-2, Development of advanced satellite based products for drought monitoring: the Vegetation Productivity Indicator (VPI). Presentation can be found at: www.oosa.unvienna.org/SAP/stdm/2002_africa/presentations/session03A/speaker02/

Early Warning System for Land Cover Change – Key to Sustainable Development and Food Security in the Qua Iboe River Basin, Nigeria

Robert Etim EKPENYONG

Summary

This paper addresses the need for a land use/land cover change early warning system. A GIS database that can be used for early warning is developed. As part of the database, land use/land cover maps of the area for two time periods (1986 and 2003) are produced.

Landuse/landcover changes in the river basin is modeled using the database. The implications of observed changes in landuse/landcover on sustainable agriculture/food security in the area are examined. The result of analysis reveals the fact that some urban centers have expanded into farmlands/fallow lands and surrounding secondary forest; mangrove forest in the area has reduced in size by 50% within a period of 17 years between 1986 and 2003; other forest covers in the area have also changed. These changes are a threat to food security among other things.

1 Introduction

The world over, early warning systems are being put in place or are already in place for natural disasters such as earthquake, volcanic eruption, hurricanes etc. The same cannot be said of the one that can warn against the negative consequences of changes in landuse and landcover.

Studies have shown that information on landuse/landcover is required for rational and sustainable allocation of land resources for development (YANG et al., 2003; EL-RAEY et al., 2000; BRONSVELD et al., 1994).

In the Qua Iboe River Basin, Nigeria, growing population densities, urbanization and poverty is leading to widespread changes in landuse and landcover. This has given rise to such problems as gully erosion, landslide, flooding, air and water pollution, degradation/ depletion of land with high agricultural potentials etc. The situation is so serious that food security and socio-economic development of many areas are being threatened.

To effectively address these problems and ensure that future generations can enjoy the benefits of the earth's resources, there is need for an early warning system for landcover change analysis and mapping. This is to among other things enhance institutional capacity to guide and control the use of and interaction with land.

2 The study area

The Qua Iboe River basin is located within Akwa Ibom State which is situated in South Eastern Nigeria. It lies between latitude 40300 and 50300N and longitude 70300 and 80150E (Fig. 1). The catchment area of the basin lies between the Imo and Cross Rivers and covers about 3266sq.km.

The climate of the basin and its environs is influenced by the north-south movement of a zone of discontinuity between maritime (Atlantic) and dry continental (Sahara) air masses. The regular movement of these air masses creates distinct dry and net seasons in the area.

The rainy season (April to October) has a greater percentage of the annual precipitation in the South. The dry season lasts from December, to February through out the area; November and March are transition months.

Mean annual rainfall over the area decreases gradually from about 4050 mm near the coastal area (the Southern part) to about 2100 mm in the North (Fig. 2).

The Qua Iboe River basin is very rich in petroleum, gas and fishery resources.

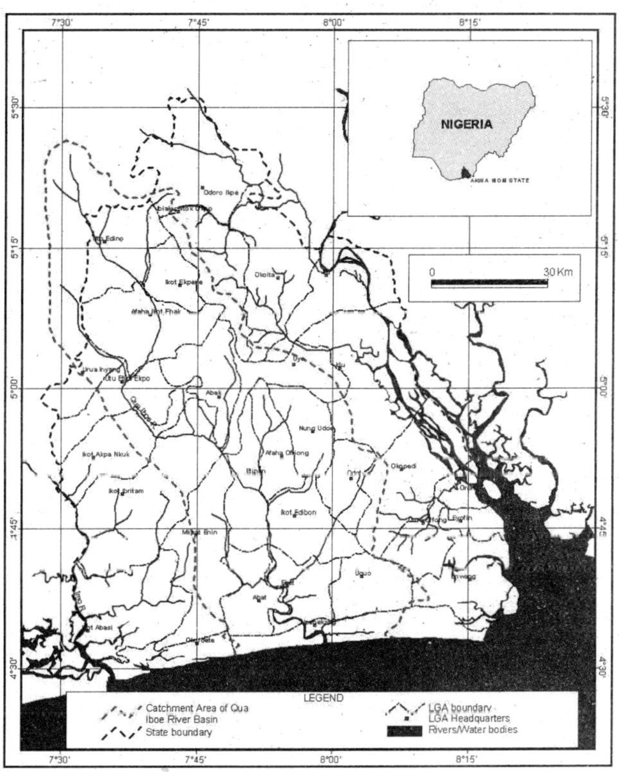

Fig. 1: Location of the study area

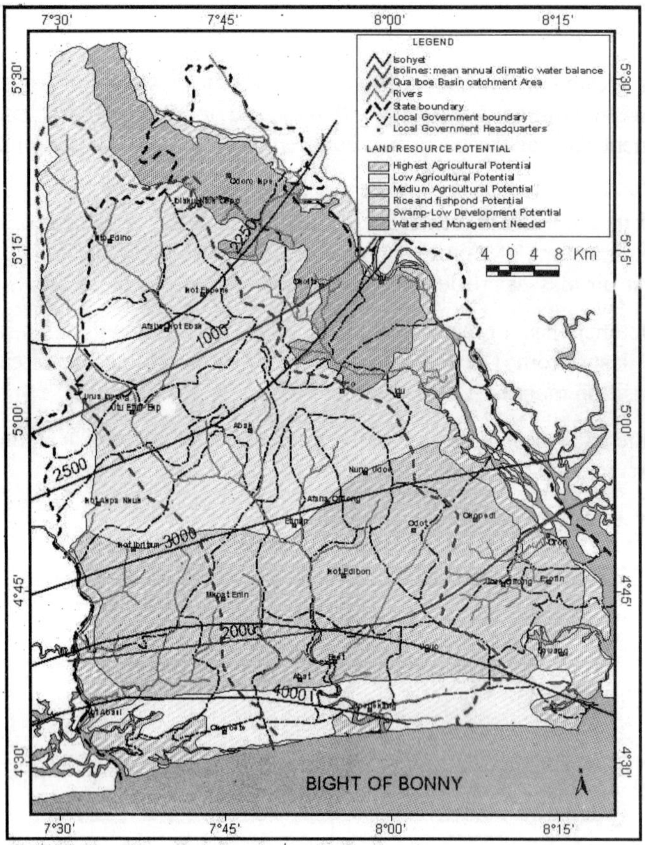

SOURCE: Cross River Basin Development Authority

Fig. 2: Qua Iboe River Basin showing rainfall distribution, mean annual climatic water balance and agricultural potential

3 Conceptual framework

Global environmental change and food systems (GECAFS) is a comprehensive programme of interdisciplinary research focused on understanding the links between food security and global change. Global environmental change, food systems and food security are key terms for GECAFS.

Global environmental change (GEC) includes changes in the physical and biogeochemical environment either caused naturally or influenced by human activities such as deforestation, urbanization, land reclamation and waste production. GEC includes changes in land cover and soils, atmospheric composition, climate variability, biodiversity etc.

Food systems are a set of dynamic interactions between and within the bio-geographical and human environments that result in the production, processing, distribution, preparation and consumption of food. The components of food system with their main elements are as follows:

- Food availability-production, distribution, exchange
- Food access-affordability, allocation, preference
- Food utilization-nutritional value, social value, food safety

Food security is underpinned by the food system. Food security is the state achieved when food systems operate such that 'all people, at all times, have physical and economic access to sufficient, safe and nutritious food to meet their dietary needs and food preferences for an active and healthy life' (FAO, 1996). Food security diminishes when food systems are stressed (GECAFS, 2005).

It is obvious from the foregoing that, one of the most important components of the food system and by extension food security is food availability. Among the elements associated with food availability, production is the most important.

The assumption here is that, if food is not produced, we cannot talk of distribution, exchange and all the elements associated with food access and food utilization.

Furthermore, studies have shown that, the production of food in any area depends on many factors. Of these factors, land is the most valuable. This is because, today, man has been able to counter unfavourable weather/climatic conditions through irrigation and building of greenhouses (BRONSVELD et al., 1994, LILLESAND and KIEFFER, 1994, UBOLDI and CHUVIECO, 1997).

In the case of land, we can only ensure availability by effecting changes in the use to which it is put. Even then, once an agricultural land is converted to uses other than agriculture, it becomes difficult to convert it back to agricultural use. This explains why the landuse/land cover change early warning system is important especially for places like the Qua Iboe River Basin where land with high agricultural potential is limited.

Furthermore, landcover change monitoring requires high spatial and temporal resolution satellite data. The 30-m-resolution Landsat has become the standard (LILLESAND and KIEFFER, 1994).

4 Materials and methods

4.1 Data

Maps of the area, mainly those showing relief and drainage, soils, vegetation, rainfall and temperature distribution, land capability/suitability, land resource development areas etc., published in 1982 by Cross River basin Development Authority at a scale of 1:250 000 were scanned, geo-referenced and digitized to create the early warning system baseline dataset.

Landuse/landcover maps were established for the year 1986 and 2003 using Landsat TM satellite data. Supervised classification was carried out to produce landuse/landcover maps for the different periods. Random test fields were located in the original image and in the field to perform accuracy assessment. An overall accuracy of 82.11% was estimated.

For change detection, the classified images were compared at pixel level. Using this approach, the location, nature and direction of land cover changes were identified and modelled (YANG and LO, 2002; FUNG and ZHANG, 1989; ABDELLAOUI, 1997; SINGH, 1999; MONGKOLSAWAT and THIRANGON, 1990).

4.2 The early warning system

In a nutshell, the landuse/landcover early warning system comprised a personal computer system and a database. It is located in the Department of Geography and Regional Planning, University of Uyo, AKS-Nigeria where it is used for capacity building. To enhance its capabilities, the system is frequently updated as new methodologies, models and/or data become available. Also advantage is taken of the free GIS/Image processing software and Landsat data from the internet.

According to EWC3 (2006), a complete and effective Early Warning System comprises a chain of four elements:

- Risk knowledge – prior knowledge of the likely risk scenarios communities are faced with;
- Monitoring and warning services-monitoring capabilities for these risks and rapid and reliable decision mechanisms for early warning;
- Communication – dissemination of understandable warnings to those at risk;
- Response capability – knowledge and preparedness capacity to act by all partners of the information chain.

The landuse/landcover early warning system is capable of providing information on the likely risk scenario that communities may face but the challenge now is to build partnerships that can support its monitoring, communication and response capabilities.

5 Manipulation and analysis of data for early warning purpose

Studies carried out in 1982 reveal the fact that the lower part of the basin has high agricultural potentials (Fig. 2).

With this in view, the landcover maps of the lower Qua Iboe River Basin for 1986 and 2003 were produced using Landsat TM (Fig. 3 and Fig. 4). By visual interpretation, these maps reveal the fact that between 1986 and 2003, land under cultivation reduced in size. This partly explains why over the years, foodstuff consumed in the area has continued to be imported from other states (EKPENYONG, 2003).

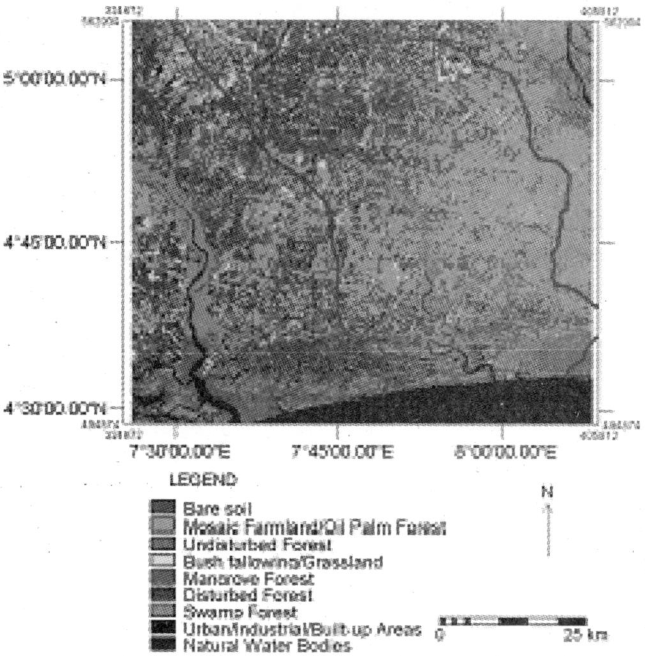

Fig. 3: Lower Qua Iboe River basin and environs: Landcover (1986)

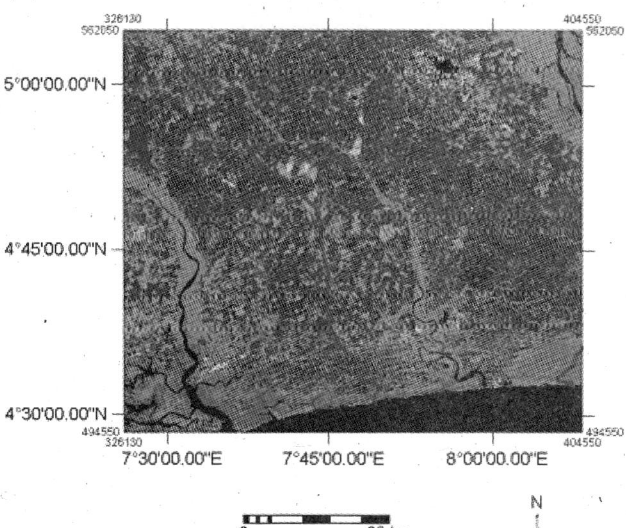

Fig. 4: Lower Qua Iboe River basin and environs: Landcover (2003)

5.1 Impact of urbanisation

Urbanisation refers to the process in which an increasing proportion of an entire population lives in urban and suburban areas. Historically, this has been closely connected with industrialisation and the discovery of petroleum in Nigeria. New job opportunities in urban areas spurred the mass movement of surplus population away from the rural communities (YANG and LO, 2003; EL-RAEY et al., 2000).

In Nigeria and indeed the Qua Iboe River Basin, the creation of new states and Local Government Areas brought about the growth in number of urban centres and urbanisation.

In Akwa Ibom state where the Qua Iboe river basin is located, there were only 10 urban centres as at 1987 when the state was created. Of this number, five were within the river basin. With the creation of new Local Government Areas, the number of urban centres in the State increased to 31. Of this number, 16 are within the catchment of the Qua Iboe River basin.

These urban centres have expanded in size as a result of urbanisation (Fig. 5 to Fig. 8). The population of people attracted to these urban centres continues to increase because of the important industries and socio-economic activity centres located within and or close to them. For instance, Uyo is a capital city and the seat of the Government of the Akwa Ibom State. Eket on the other hand is located near the operational base of EXXONMobil and SHELL petroleum.

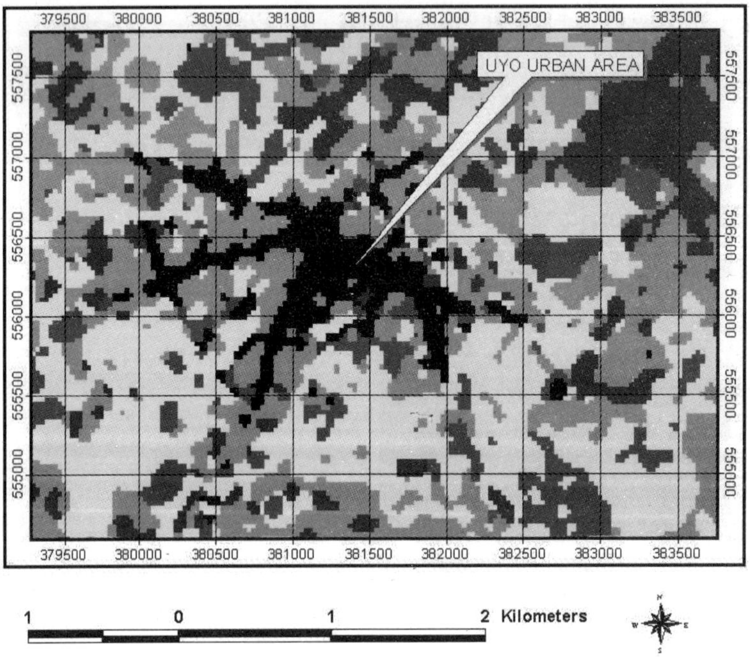

Fig. 5: Built-up area of Uyo urbans at 1986 (area = 117.05 ha)

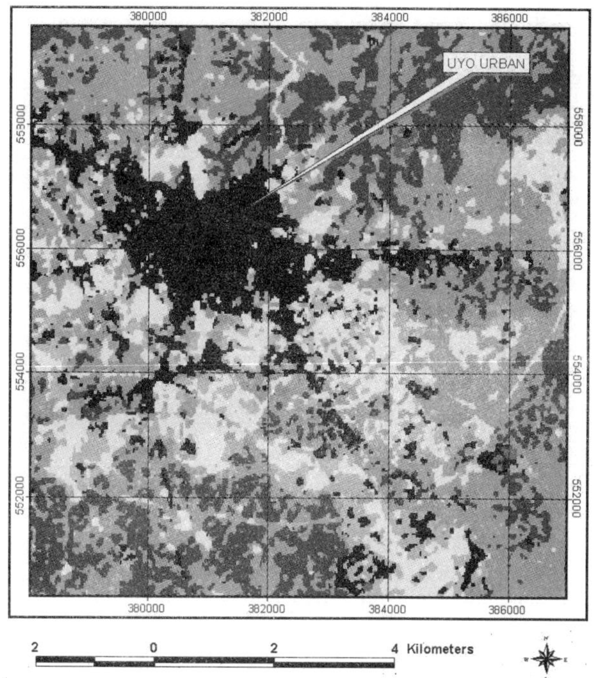

Fig. 6: Built-up area of Uyo urbans at 2003 (area = 1 472.67 ha)

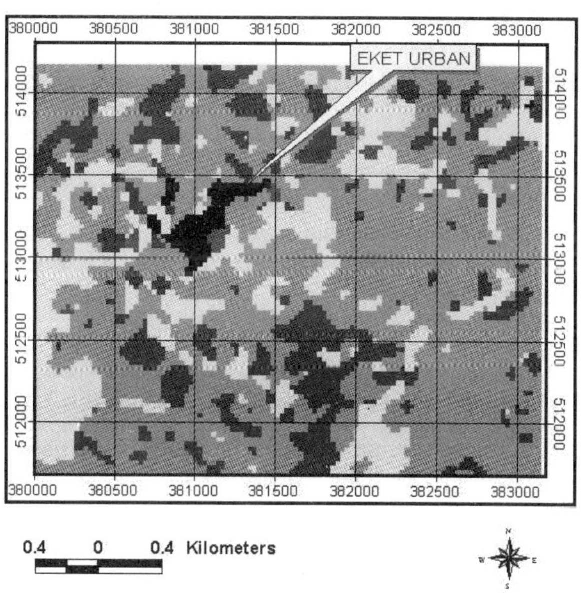

Fig. 7: Built-up area of Eket urbans at 1986 (area = 14.54 ha)

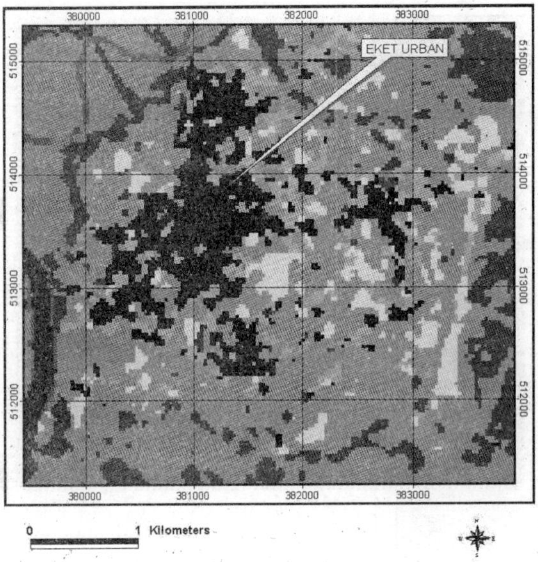

Fig. 8: Built-up area of Eket urbans at 2003 (area = 236.25 ha)

A critical visual interpretation of the 1986 images of Uyo and Eket (Fig. 5 and Fig. 7 respectively) shows that the urban areas were surrounded by agricultural land (i.e. mosaic farmland/oil palm forest and bush fallowing/grassland land cover types).

This implies that the expansion of urban areas in 2003 has encroached on valuable agricultural land. This is a serious threat to food security and sustainable agriculture in the area under study. More so, the land with high agricultural potential in the area is limited in extent (Fig. 2).

5.2 Impact of poverty

A recent study reveals the fact that poverty is widespread and severe in Akwa Ibom state and indeed the Qua Iboe River basin. About 57 percent of the people are living in poverty, about 42 percent are non-poor. Among the poor, 28 percent are classified as core or extreme poor and about 29 percent are moderately poor (EKPO and UWATT, 2005).

All the computations were carried out using measures and approaches recommended by World Bank (WORLD BANK, 2001). Details of the result of the study are given in Table 1.

The values for the Local Government Areas which form part of the Qua Iboe River Basin Catchment are high.

Most of the people living within the Qua Iboe river basin and environs are poor (Fig. 9). Poverty is clearly a major factor responsible for landuse/landcover change within and around the study area.

Table 1: Poverty incidence (head count), Depth and Severity by Local Government Areas

Local Government	Headcount (P0)		Depth (P1)		Severity (P2)	
Areas	Index	%	Index	%	Index	%
		Contribution		Contribution		Contribution
Abak	41.33	2.47	17.27	2.06	9.39	1.67
Eastern Obolo	66.67	2.93	14.85	3.67	31.08	4.06
*Eket	46.48	2.01	13.72	1.19	6.68	0.86
*Esit Eket	45.70	2.10	16.79	1.55	8.47	1.16
*Essien Udim	30.81	1.62	14.52	1.52	8.91	1.39
*Etim Ekpo	75.13	4.33	45.69	5.26	31.93	5.48
*Etinan	39.79	2.32	11.97	1.39	5.35	0.93
*Ibeno	30.96	1.86	7.00	0.84	2.67	0.48
*Ibesikpo Asutan	69.35	3.93	36.54	4.14	23.17	3.91
Ibiono Ibom	66.67	4.02	29.80	3.60	17.52	3.15
*Ika	53.37	3.14	14.95	1.76	5.99	1.05
*Ikono	49.74	2.96	17.73	2.11	9.26	1.64
Ikot Abasi	43.98	2.56	19.66	2.29	12.48	2.16
*Ikot Ekpene	78.31	4.51	39.50	4.55	26.78	4.60
Ini	83.64	5.06	37.78	4.56	23.53	4.23
Itu	59.56	3.32	26.40	2.94	15.62	2.60
Mbo	68.75	4.02	37.93	4.44	26.44	4.61
*Mkpat Enin	56.18	3.05	31.33	3.40	22.63	3.66
*Nsit Atai	49.23	2.93	24.78	2.95	16.22	2.87
*Nsit Ibom	77.04	4.60	55.00	6.57	41.38	7.36
*Nsit Ubium	59.38	3.47	15.84	1.86	6.19	1.08
*Obot Akara	50.82	2.83	23.95	2.67	15.43	2.56
*Okobo	90.36	5.43	77.91	9.35	69.10	12.36
*Onna	45.60	2.53	17.35	1.92	9.46	1.56
Oron	71.36	4.33	39.72	4.82	26.96	4.87
*Oruk Anam	45.93	2.41	13.27	1.39	5.94	0.93
Udung Uko	77.89	4.72	57.16	6.93	46.44	8.39
*Ukanafun	26.88	1.52	10.54	1.20	6.58	1.11
*Uruan	36.56	2.07	10.70	1.20	4.70	0.79
Urueoffong Oruko	67.20	3.87	44.25	5.10	32.58	5.59
*Uyo	41.21	2.50	13.24	1.61	6.29	1.14
Overall Poverty (%)	56.77		28.39		19.06	
Core Poverty (%)	27.51		14.24		9.24	

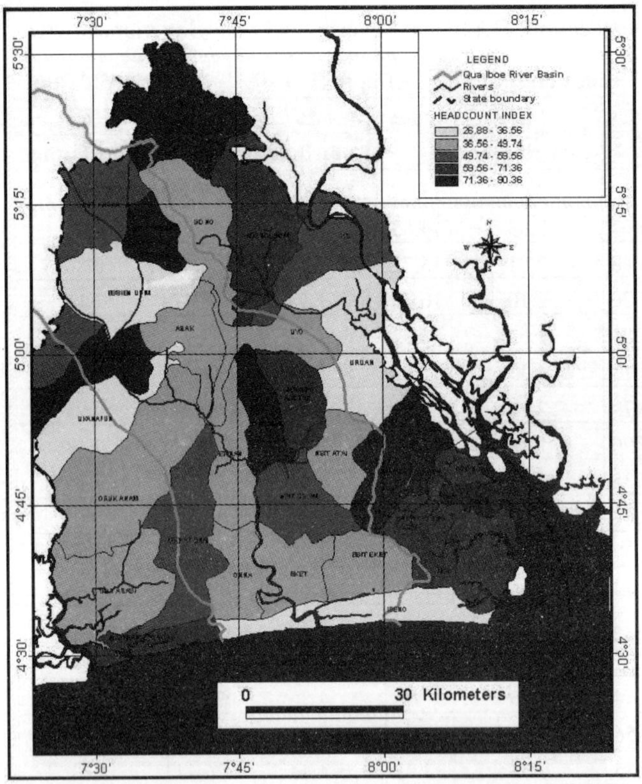

Fig. 9: Map showing the spatial distribution of poverty in the Qua Iboe Basin and environs

Ironically, people's effort to escape from poverty damages the environment. For example in their struggle to survive, poor people often have no choice but to destroy their surrounding by cutting down trees, overworking the soil, over grazing range lands and over fishing (CAMP, 1990).

One of the landcover changes that have taken place as a result of poverty in the area involves the mangrove forest.

Figure 10 and figure 11 reveal the fact that mangrove forests in the Qua Iboe River Basin have been reduced in size from 2 300.54 ha in 1986 to 1 248.48 ha in 2003.

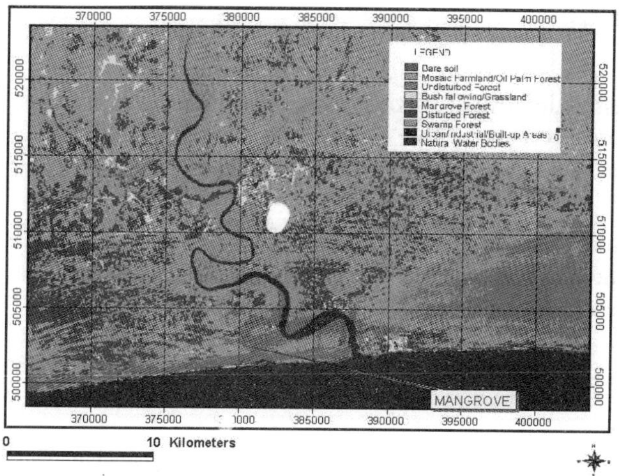

Fig. 10: Lower Qua Iboe Basin showing the distribution of mangrove forests in 1986

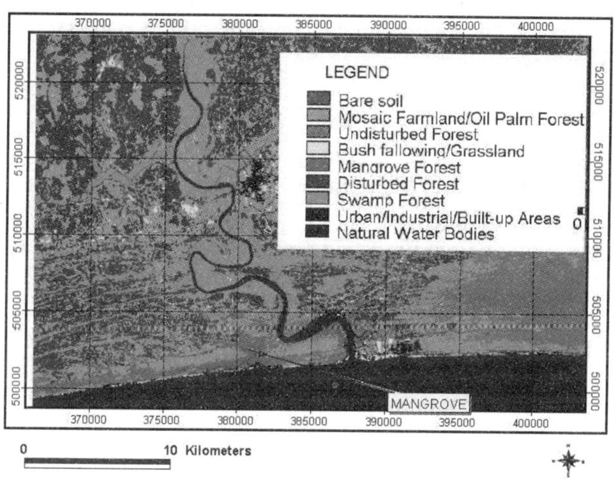

Fig. 11: Lower Qua Iboe Basin showing the distribution of mangrove forest in 2003

The consequence of this is severe. Through the action of its roots, a mangrove forest recycles nutrients and traps land based debris, sediments and suspended particulate matter carried to the coast by rivers. The forest and its roots secure the land, preventing shoreline erosion (YEQIAO et al., 2003).

The continuous decline of mangrove in many parts of the world has resulted in sedimentation problems in offshore reefs and undermined the sustainable supply of fish stocks. (SPIERS, 1999; ALEX et al., 2003).

6 Conclusions

This study confirms reports of destruction of forest cover especially the mangrove forest. Losing 50 percent of a valuable forest resource within such a short duration leaves a lot to be desired. Both the present and the future generations stand to suffer bitter consequences if this trend continues.

The results of this study attest to the fact that with the landuse/landcover change early warning system, it is possible to monitor and manage agriculture more effectively with the aim to ensure food security among other things.

References

ALEX, H., CATHERINE, T., LEO, L. and NICHOLE, W., 2003, High Resolution Mapping of Tropical Mangrove Ecosystem using Hyperspectral and Radar Remote Sensing. International Journal of Remote Sensing, 24(13), pp. 2739–2759.

AYENI, O.O., 1997: Digital Mapping and GIS Education in Developing Countries. In IKHUORIA, I.A. (Ed.), Cartography and Geographical Information Systems for Sustainable Development. Nigerian Cartographic Association, Benin City, pp. 25–36.

BRONSVELD, K., CHUTIRATTANAPAN, S., PATTANAKANOK, B., SUWANWERAKAMTORN, R. and TRAKOOLDIT, P., 1994, The Use of Local Knowledge in Land Use/Land Cover Mapping From Satellite Images. ITC Journal, 2, pp. 349–358.

BRONSVELD, K., HUIZING, H. and OMAKUPT, M., 1994, Improving Land Evaluation and Land Use Planning. ITC Journal, 2, 359–365.

CAMP, S.I., 1990, Population Pressure, Poverty and the Environment. Lecture presented at the conference on Endangered Earth: An Evolutionary Perspective. University of California.

EWC3, 2006, Third International Conference On Early Warning: From Concept To Action. www.ewc3.org.

EKPENYONG, R.E., 2000, Trade. South Eastern Nigeria: It's Environment. Kaduna, Nigeria, abeam publishing co., pp. 144–149.

EKPO, A.H. and UWATT, B.U., 2005, Akwa Ibom state of Nigeria: Poverty Profile and Economic Transformation. Uyo, Foundation for Economic Research and Training.

EL-RAEY, M., FOUDA, Y. and GAL, P., 2000, GIS for Environmental Assessment of the Impacts of Urban Encroachment on Rosetta Region, Egypt. Environmental Monitoring and Assessment, 60, pp. 217–233.

FAO, 1996, Report of the World Food Summit. Rome:FAO.

FUNG, T. and ZHANG, Q., 1989, Land Use Change Detection with Landsat Digital Data in the Kitchener-Waterloo Area. Remote Sensing and Methodologies of Land Use Change Analysis. University of Waterloo, Department of Geography Publications Series, Occasional Paper 6, pp. 135–153.

GECAFS, 2005, Science Plan and Implementation Strategy. Earth System Science Partnership [IGBP,IHDP,WCRP,DIVERSITAS]. Wallingford, Report 2, pp. 36.

JENSEN, J., 1995, Introductory Digital Image Processing: A Remote Sensing Perspective. Prentice Hall, New Jersey.

LILLESAND, T.M. and KIEFFER, R.W., 1994, Remote Sensing and Image Interpretation. John Wiley and Sons, NewYork.

MONGKOLSAWAT, C. and THIRANGOON, P., 1990, Land Cover Change Detection Using Digital Analysis of Remotely Sensed Satellite Data: A Methodological Study. Asian Conference of Remote Sensing (ACRS).
http://www.gisdevelopment.com/aars/acrs/1990/G/lclu003.shtml

SINGH, A., 1989, Digital Change Detection Techniques Using Remotely-Sensed Data. Int. J. Remote Sensing, 10(6), pp. 989–1003.

SPIERS, A., 1999, Review of International Continental Wetland Resources. In: FINLAYSON, C. and SPIERS, A. (Ed.), Global View of Wetland Resources and Priorities for Wetland Inventory. Canberra, Australia, Supervising Scientist Report 144.

UBOLDI, J. and CHUVIECO, E., 1997, Using Remote Sensing and GIS to Assess Current Land Management in the Valley Of Colorado River, Agentina. ITC Journal, 2.

YANG, X. and LO, C., 2002, Using Satellite Imagery to Detect Land Use and Land Cover Changes in Atlanta, Georgia Metropolitan Area. Int. J. Remote Sensing, 23(9), pp. 1775–1798.

YANG, X. and LO, C., 2003, Modeling Urban Growth and Landscape Changes in the Atlanta Metropolitan Area. International Journal of Geographical Information Science, 17(5), pp. 463–488.

YEQIAO, W., GREGORY, B., JARUNEE, N., MICHAEL, T., AMANI, N., JAMES, T., LYNNE, H., ROBERT, B. and VEDAST, M., 2003, Remote Sensing of Mangrove Change along the Tanzania Coast. Journal of Marine Geodesy, 26, pp. 35–48.

WORLD BANK, 2001, World Development Report: Attacking Poverty. Oxford University Press, New York.

Mapping ODAdata – A Web Based Mapping Toolset to Explore Complementarities and Coherence in International Cooperation

Alexander BOHR and Frank GOTTSMANN

Summary

Responding to the urgent need and recent political decisions at highest level (Paris HLF II, EU Council Decision on EU Aid Effectiveness Package) for closer coordination & harmonization between donors as well as a lack of tools rendering Aid Flow Transparency, ODAdata with its modules ODAMoz and ODAMap, provides for a unique standardized and cost effective solution, easily trar ferable to any other developing country. The transparency of *Aid Flows* and the dissemination of cooperation project information is the fundamental basis for more effective formulation, management and implementation of development policies and programmes. Managers, stakeholders and decision-makers need such comprehensive information in a user-friendly way giving them, at a glance, an overview on past, running and future projects with their demands, problems and impacts as well as their various direct and indirect inter-relationships. With such information, decision-makers can seek a wise balance between the many conflicting demands being made on ODA (Official Development Assistance) resources ensuring that available funds are well invested and its capacity are not exceeded. With the first successful implementation now taking place in Mozambique, several countries are starting to become interested in applying and implementing a similar tool.

1 Introduction

Primary objective and focus of ODAmoz is to provide donors with a toolset to fulfil their obligations to report and forecast on disbursements towards the Government of Mozambique. Donors and the Government of Mozambique will thus be provided with a monitoring tool for Aid Effectiveness (e.g. according to indicators established by OECD/DAC members in the Paris High Level Forum II meeting in 2005) and hence gain a better understanding of activities of individual donors in the country.

ODAmoz (ODAdata toolset and services for Mozambique) was developed to capture information on cooperation projects and programs, with criteria such as sector, donor, location, implementing agency, duration, funding, MDG goals etc, of all donor agencies inclusively UN participation. Thus ODAmoz has become a central electronic tool for donor coordination and harmonization accessible by all comfortably through the internet: http://www.odamoz.org.mz/reports/rpt_default.asp.

The ODAdata toolset was implemented through funding from the European Commission as part of the EU Pilot Initiative on Coordination & Harmonization of community policies and thus adds significant value to the harmonization of the development cooperation work of all EU Member States and other donors, members of the Donor Partners Group (DPG), providing ODA to Mozambique.

2 ODAmoz

The implementation of the ODAdata toolset has its origins in several decisions by the EU to strengthen coordination & harmonization in order to reach a higher degree of Aid Effectiveness.

In March 2002, in preparation for the Monterrey Conference, the Council of the European Union agreed in Barcelona to take concrete steps in the area of co-ordination of policies and harmonization of procedures before 2004. This commitment was reinforced through the insertion of a similar element into the Monterrey Consensus. EU efforts are to be seen in the context of global efforts involving donors and partner countries worldwide, led by OECD-DAC. The importance of the development partners' drive for harmonization is encapsulated in a declaration made at the end of a high-level forum, HLF I, which took place in Rome in February 2003. This states, inter alia: 'Our deliberations are an important international effort to harmonize the operational policies, procedures, and practices of our institutions with those of partner country systems to improve the effectiveness of development assistance and thereby contribute to meeting the Millennium Development Goals (MDGs).'

In June 2002, in order to prepare the implementation of its political commitment to greater co-ordination and harmonization, EU Member States decided to ask the Commission to launch a pilot initiative. The mission analyzed four country cases: Morocco, Nicaragua, Mozambique and Vietnam.

The conclusions of the pilot initiative in 2002, which recognised that a certain amount had already been achieved in the context of co-ordination and harmonization, but that much more needed to be done, led the EU Directors General for Development to agree that in addition to a global action plan for harmonisation and co-ordination being drawn up, the country-specific action plans, Road Maps, for the four pilot countries should be proposed jointly by the EU (Member States and EC) delegates in the field and presented as the result of the efforts of the local community of EU donors.

In 2004, as a follow up to the pilot initiative in 2002, an EU Donor Atlas with special focus on the four country cases: Morocco, Nicaragua, Mozambique and Vietnam, was produced. Since then the EU Member States have started a debate on complementarities using the Atlas as one of the tools available (currently, the EU Donor Atlas has been updated and a regional issue for West Africa is in production).

In mid 2004, the EU Heads of Cooperation in Mozambique reacted positively to start work on the EU Projects Database (DbIS, the ODAmoz predecessor) proposed by the EC Delegation, accessible to all EU Member States and the Government, which could be the basis for information sharing and coordination. So far, information on Projects and financial disbursements had been sent, if at all, to the Government of Mozambique in

EXCEL format. Data was neither standardized nor was the information shared between donors.

In January 2005, DbIS implementation was started through financing from the European Commission as part of the EU Pilot Initiative on Coordination & Harmonization of community policies and later continued through the EDF Country Programme. The political mandate for EU Donor participation came from an agreed EU Road Map for Mozambique on Coordination and Harmonization which called for a set of actions including the establishment of an EU Online Projects Database.

As the first reporting on quarterly disbursements towards the Government of Mozambique for the 2nd Quarter 2005 was successful, the proposed format and design of DbIS was accepted and seen as an impressive way to manage ODAdata for Mozambique in line with OECD DAC definitions and the just concluded seconded High level Forum on Aid Effectiveness in Paris.

Due to the harmonization process in Mozambique being far ahead and exemplary for developing countries, especially in reference to the Programme Aid Partners to Mozambique Budget Support Group, PAPG18, it was decided to open the database to all non-EU donors as well: EU+ DbIS. By January 2006, USAID, Japan, Norway, Switzerland, the World Bank and the African Development Bank where integrated into EU+ DbIS.

Having successfully included the majority of donors present in Mozambique (80–90 percent of all ODA to Mozambique covered, or some 1 000 000 000 US $ disbursements for 2005), first analyses could be started next to the drafting of the full EU+ DbIS 2005 report. The drafting of a Donor Atlas on country level was seen as most promising, using the geographical information available in EU+ DbIS. The results of the analyses, showing the spread of donor disbursements per province and sectors or the amount of donors active per sector and province, proved to be astonishing, especially when linked to the poverty reduction strategy:

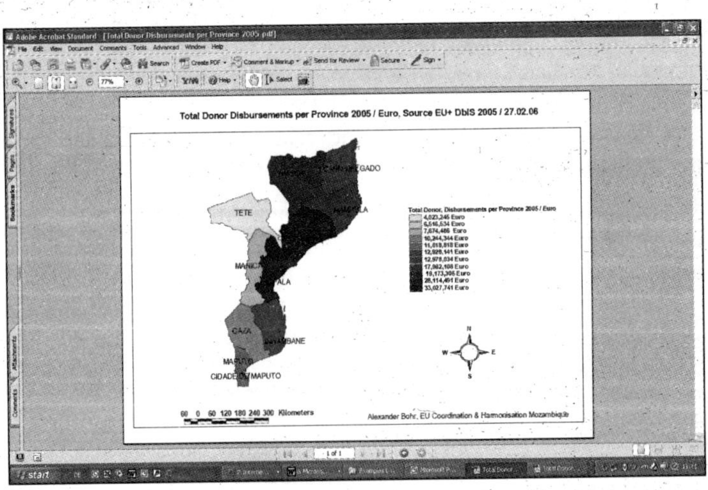

Fig. 1: Geographical distribution per province of Donor Aid Flows in Mozambique, based on EU+ DbIS 2005

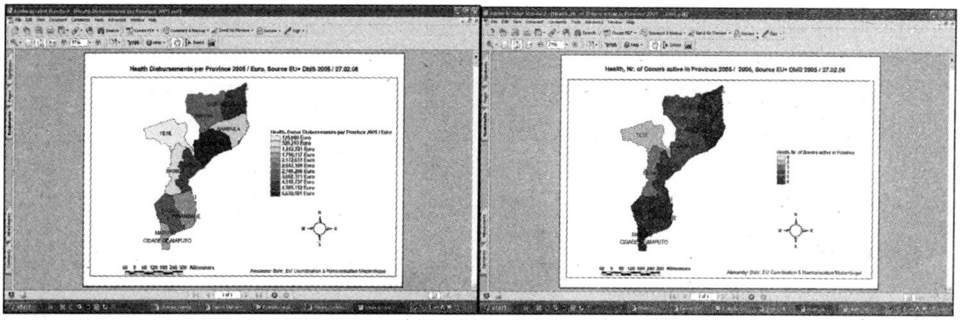

Fig. 2-3: Geographical distribution per Province and Health sector, numbers of donors active per province in the Health Sector, based on EU+ DbIS 2005

For data entry and maintenance, system participants like donor agencies select a focal point that (after having received necessary training) is in charge of autonomously collecting data, entering and up-dating it on the ODAmoz project database via the web on a quarterly basis. The stored encompassing datasets lay ground for information from which the systems users will be able to design own reports in the form of tables and geographic maps using defined criteria according to specific needs. As ODAmoz has been growing and becoming more and more donor inclusive, it is seen to be a very useful tool e.g. in the On-/Off-Budget Public Finance Management discussions with the Government of Mozambique and in providing a common source of information for participating donors that are revising their country strategies in 2006.

In March 2006, in order to complete the picture of donors present in EU+ DbIS, it was decided to merge the EU+ DbIS database with the UN projects database, the new name becoming ODAmoz, Overseas Development Aid to Mozambique Database:

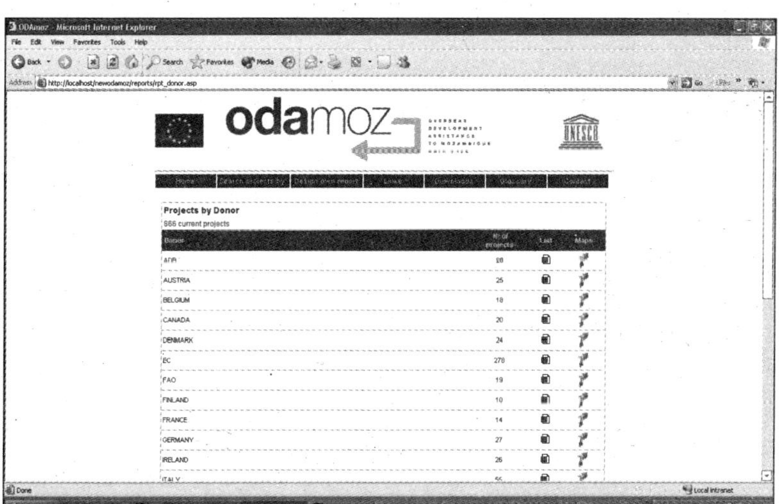

Fig. 4: View of the ODAmoz online database predefined search per donor, http://www.odamoz.org.mz

Due to the success of the first Draft Donor Atlas for Mozambique and the lack of available geo-processed donor data, it was further decided to integrate a Web mapping module into the ODAmoz structure, ODAmap, which will replace the current map image display and caters for the future use of GIS functionalities for online Donor Atlas style queries and data mining.

3 ODAmap

Next to simple tables and graphs, the ODA project relevant information will be made available at a glance, by the web-map based Information System ODAmap. The UMN-MapServer software, originally developed by the University of Minnesota in cooperation with NASA and the Minnesota Department of Natural Resources was adapted to the requirements of the already successfully running EU+ information system DBIS (ODAmoz ancestor) as an Open Source development environment to support it with spatially enabled Internet applications (http://mapserver.gis.umn.edu/ index_html?set_language=en).

The ODAmap module builds on an architecture consisting of three components, interacting closely together. The three components of the mapping module are (a) the mapserver itself, (b) a PostgreSQL RDBMS and (c) an Apache webserver. The mapserver draws maps based on vector and raster data model drawing on geo-data stored in the RDBMS, thus complicated and permanent hand-drawing of images (digitisation) to visualise a certain database content (e.g. which donor is active in which province?) is taken over by the UMN. Thanks to its PostGIS extension, the special feature of the PostgreSQL RDBMS is its ability to store spatial (e.g. the boundaries of Mozambican provinces & districts) and non-spatial data together coherently in one database. This simplifies the administration of all ODA-project relevant data enormously and prevents redundancies. Furthermore MapServer provides sufficient functionality to support a wide variety of web applications: beyond browsing GIS data, it allows to create geographic image maps, i.e. maps that can direct users to content.

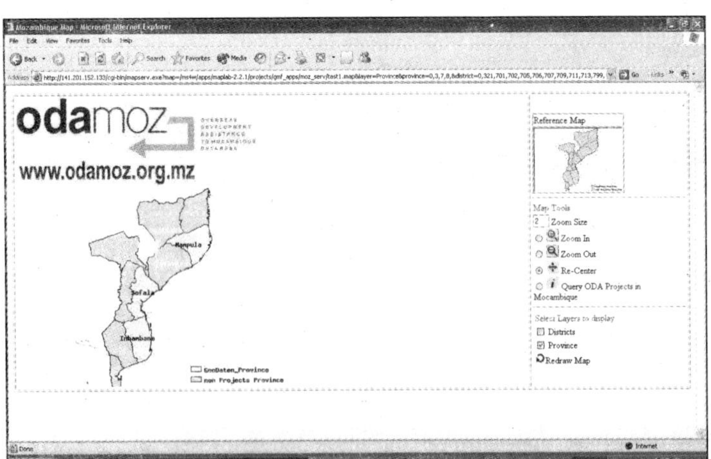

Fig. 5: View of the draft ODAmap, showing the provinces in which Austria is active

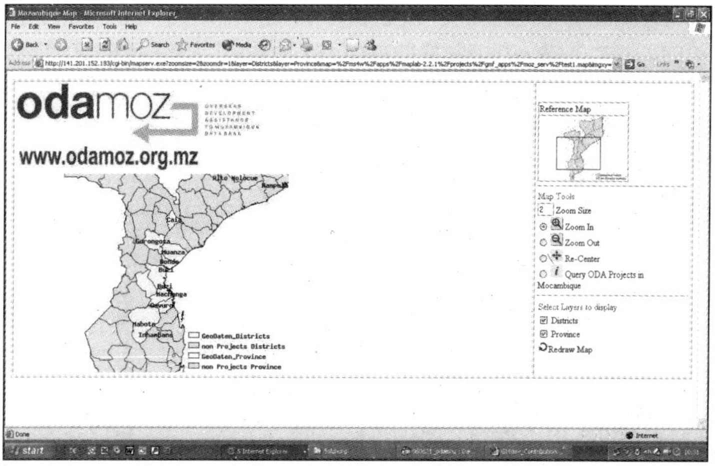

Fig. 6: View of the draft ODAmap, showing the districts in which Austria is active

Technically, the MapServer generally runs as a Common Gateway Interface (CGI) application from a http-server. The ODAdata CGI application thereby uses the following resources:

- An Open Source Apache or Microsoft Internet Information Server (IIS).
- An initialization file to start the MapServer software to trigger the first view of the MapServer application.
- A Mapfile to control the MapServers' use of data, in an application. This includes display and query parameters respectively.
- A template file that controls the MapServer application's user interface in the browser window. The template file that allows the author to place the map and legend on the http-page, and determines the interaction with the MapServer application like browsing, querying, zooming etc.
- A spatial dataset stored in PostgreSQL and managed via PostGIS Extension. Many open and proprietary file and geo-data formats are accepted by the UMN, among others the ESRI shapefile format and various Raster data formats (GeoTiff, Tiff etc.).

The whole ODAdata system is set-up on a UNIX server allowing for traffic volume rates up to 100 GB.

4 Outlook

In order to avoid the creation of similar but independent and non-compatible individual country solutions, it is suggested that ODA data tools as well as the ODAdata structure will be adapted to provide for the possible future extension to other countries on a common standardized open source platform, ideally allowing national, regional as well as global ODA data mining. With the ODAmap module, the ODA-analysis and reports will not any longer be limited to simple database queries: *Who is doing what with how much money?*

Moreover, the information capabilities can be extended to the spatial dimension by allowing to ask: *Who is doing what with how much money and where*? – and how close and embedded is it to other activities and infrastructure in the region?

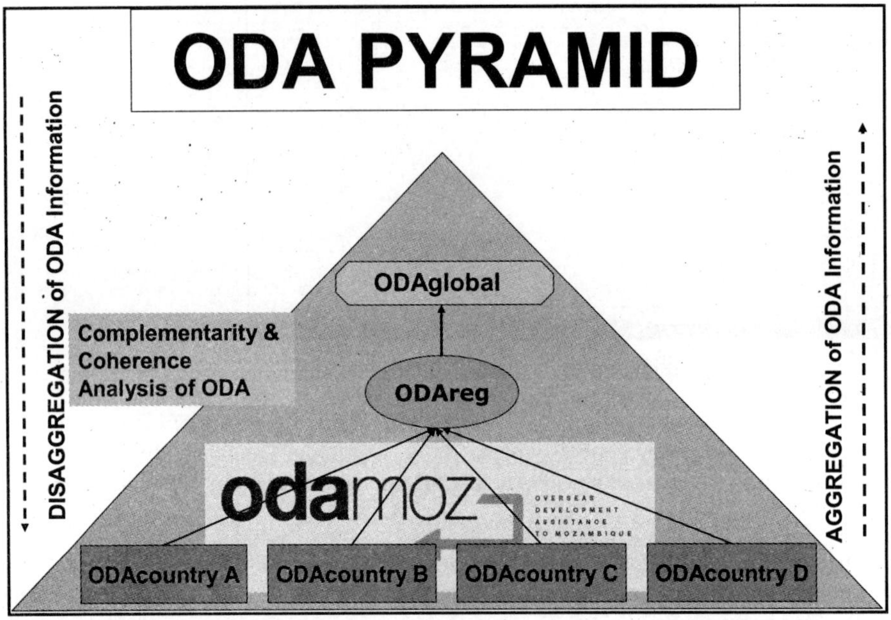

Fig. 7: Proposed implementation of ODAdata in multiple countries

References

UMN Mapserver:
http://mapserver.gis.umn.edu/index_html?set_language=en.

The European Community's Development Policy, Statement by the Council and the commission:
http://ec.europa.eu/comm/development/body/legislation/docs/council_statement.pdf#zoom=100

The Paris Declaration on Aid Effectiveness:
http://www.oecd.org/dataoecd/11/41/34428351.pdf

Chapter 4

Contributions from Panel Members and Session Chairs

With the objective to discuss opportunities, achievements and challenges for the development of partnerships and the realisation of cooperation projects and programmes, the organisers invited experts with relevant backgrounds from executing organisations as members of a panel concluding the conference. The following contributions reflect the main issues raised during their opening statements and provide aspects from different perspectives:

Ges. Mag. Franz Hörlberger, Austria
Austrian Ministry of Foreign Affairs
Former Head of Department VII.3
Humanitarian Assistance, NGO-Co-financing and Business Development
Now: Ambassador of the Republic of Austria, Hashemite Kingdom of Jordan

Dr. Juan Carlos Villagrán de León, Guatemala
Director: Centro de Investigación y Mitigación de Desastres Naturales CIMDEN-VILLATEK, Guatemala, and
United Nation University, Institute for Environment and Human Security, Germany

Prof. E. Alaphia Wright, Sierra Leone
Former Dean of Engineering, University of Zimbabwe and
Internal Oversight Service UNESCO, France

David Stevens, Brazil
United Nation Office for Outer Space Affairs
Space Application Programme
UN Vienna, Austria

Two session chairs provided additional input by reflecting on the theme of the meeting in the context of their work:

Dr. Dusan Sakulski, South Africa/Serbia
Former System Coordinator at the National Disaster Management Center, South Africa
Now: United Nations University, Institute for Environment and Human Security, Germany

Dr. Tsehaie Woldai, Eritrea
President, African Association of Remote Sensing of the Environment (AARSE) and
Associate Professor Geological Remote Sensing, Portfolio Manager Education
Department of Earth Systems Analysis
International Institute for Geo-Information Science and Earth Observation (ITC), the Netherlands

The range of topics covered spread from North-South cooperation in science and technology, GIS for risk assessment in Guatemala, to geoinformation for poverty alleviation and a general reflection on geoinformation, sustainable development and decision-making.

The last two papers report on the status of the development of a national risk assessment system in South Africa, and the existing infrastructure and resources for remote sensing applications on the African continent.

Geoinformation (GI) for Development in the Wider Context of North-South Scientific and Technological Cooperation

Franz HÖRLBERGER

1 Introductory remarks

In the current 'Three-Year Programme on Austrian Development Policy' – the strategic guidelines for development interventions – scientific and technological North-South cooperation is not mentioned explicitly as a priority instrument as development cooperation is generally understood to focus on educational cooperation in the wider sense. However, the concept for educational cooperation as a sector policy is currently under review.

The primary objectives in the 3-year Programme are stated as:

- reducing poverty,
- preventing conflicts and
- protecting the environment.

The question is in how far GI – as part of science & research – can help to achieve these objectives? I do believe that GI can indirectly contribute to the above objectives.

Austria, like all the other UN member states, has adopted the outcome of the 2005 World Summit for Sustainable Development (WSSD Programme of Action). Under point 60 the vital role that science and technology play for the achievement of the development goals is recognized. Under this – and other – international reference frameworks the UN member states commit themselves to a number of concrete measures, among others:

- to support R&D, including through voluntary partnerships …
- to facilitate access to and the development, transfer and diffusion of technologies …
- to harness new agricultural technologies in order to increase productivity …

2 Science and technology/research and Austria's development cooperation

In Austria various forms of research cooperation exist. Several federal ministries – such as BMBWK (Education, Science and Culture), BMVIT (Transport, Innovation and Technology), BMF (Finance), BMaA (Foreign Affairs), and scientific / academic institutions engage on their own initiative in development policy / development projects e.g. by the Commission for Development Studies at the Austrian Academy of Sciences (KEF/ÖAW) in the areas of ecology, energy, agriculture. Bilateral development projects of the Austrian Development Cooperation (ADC) support research institutions or programmes in developing countries e.g. International Centre of Insect Physiology and Ecology (ICIPE), International Laboratory for Research on Animal Diseases (ILRAD).

ADC has for many years supported North-South cooperation in the field of science & research through post-graduate training programmes mainly for scientists and researchers, and through capacity building for universities and research institutes (North-South Cooperation) in developing countries. An example is the International Post-graduate Training Programme in Limnology carried out by the Institute for Limnology of the Austrian Academy of Science. Close to 500 scholarships are provided per annum for a North-South Exchange Programme mainly targeting post-graduate students. Considering the issue of brain drain, the scholarship policy could be revised in a way that third world students are supported to study in their own/neighbouring countries rather than in the Northern hemisphere.

Numerous university partnerships between universities in Austrian and in third world countries (predominately ADC cooperation countries: Nicaragua, Cape Verde, Burkina Faso, Ethiopia, Uganda, Mozambique, Bhutan

Albania, Bosnia and Herzegowina, Macedonia, Serbia und Montenegro, Moldovia, Croatia) have been established often ad hoc as a result of personal or institutional contacts as well as through the facilitation by Austrian Embassies rather than through a programmatic approach by ADC. An example here is the cooperation between BOKU (Agricultural University Vienna) and Egerton University in Kenya.

3 Comments on partnerships

Generally university partnerships work best when both sides – North/South, science and research, donors and recipients – benefit from the cooperation. Among the criteria for successful partnerships are:

- mutual benefit
- synergies
- capacity building
- equal partners
- ownerhip and inclusiveness
- cultural sensitivity

and the objectives of the research should only be set in an atmosphere of shared responsibility between the partners.

Frequently arguments used against university partnerships stress the fact that focussing on science & research comes, more often than not, at the cost of universal education, and favours primarily elites! In addition, the research themes are too narrowly oriented along the needs and objectives of developed (!) countries. And the cooperation between university partners encourages brain drain.

Shouldn't we encourage "brain circulation" rather than only be fearful of "brain drain"?

4 How can we improve the cooperation between science and technology and development or more specifically of Geoinformation in development cooperation?

We need to develop a more systematic and strategic approach instead of dealing with science & technology in development cooperation on an ad-hoc basis. For this to achieve, we need adequate human and financial resources; do we have them? Can we expect them in the near future? Rather not, if we consider current streamlining of personnel and budgetary cuts in the public sector. But let us not despair: where there is a will there is a way! Innovative thinking is required!

- We must highlight the added value of S&R, of GI for development.

We need an open, more intensive dialogue to learn more about each other and to share information. To this end, we need dedicated people on both sides to facilitate more interaction between GI and donors. As a first step let us identify areas of possible cooperation / overlap / applicability of GI for development e.g. early warning, disaster mitigation, sustainable use of natural resources, increasing agricultural productivity, environmental management / conservation, natural resource management. There is indeed a wide scope where GI can be used as an appropriate instrument for development, as can be noted from the presentations given during this programme.

- How can we develop an interface between GI and development?

We have to facilitate exchange of knowledge and experience through conferences like this one, where the application of GI is demonstrated not only amongst the GI community itself but shared and discussed with development actors!

- How can we tap the developmental potential of GI (public/private)?
- We also need to critically assess development actions whether added value is in fact created for the development process!

Duplication of efforts (e.g. data capture) has to be avoided and the coordination among development actors is called for. (See www.odamoz.org.mz, and the respective article in this volume)

Let me point out some examples:

GI for disaster mitigation:
In the Province of Sofala, Central Mozambique: GI was most effectively used during the floods in 2000 and the reconstruction phase which followed.

GI for early warning:
In Bhutan, an early warning system was established around the glacier Luggye Tsho to reduce the risk to communities in the catchment area caused by glacier lake outbursts as a result of global warming.

Information and communication technology (ICT) for development:
The digital divide is still wide and specifically Africa is lagging behind. How to adapt the potential and the content of ICT to suit e.g. Africa`s social and economic systems, with emphasis on poverty reduction and economic growth?

IT for Small and Medium Enterprises in Uganda:
Austria co-finances a UNIDO project

IT-university course at Birzeit University/Palestine

The Way Forward

The various presentations have underlined that GI and development have many points of contact. Let us encourage even more interaction between GI scientists and development planners on a theoretical as well as practical level. We, the development actors, can certainly make better use of the potential GI offers by integrating the various disciplines of GI not only into the planning and programming cycle but also in the implementation phase in the field.

Thank you for your attention.

N.B.: The content reflects the personal opinion of the author and do not represent the official development policy of the Austrian Government.

Risk Assessment Using GIS Tools – Examples from Guatemala

Juan Carlos VILLAGRÁN DE LEÓN

Summary

Risk assessment is an essential step in disaster-risk management. Risk assessment allows decision makers to understand and dimension the causes of disasters associated with natural phenomena, facilitates the allocation of resources to minimize such risks in the context of promoting sustainable development, and identifies potential needs regarding humanitarian assistance to aid communities which have faced disasters as a consequence of their high risk. This article presents examples of risk assessment spanning from the household level to the state or provincial level using GIS tools, and identifies challenges ahead if such tools are to be used for purposes of presenting visually information regarding risks associated to communities of different sizes.

1 Introduction

Guatemala is a country which is frequently experiencing disasters as a consequence of its location in hazardous areas and because vulnerabilities are continually generated throughout the country due to various social and economic reasons. Being located in the Central American isthmus, it is located within the path of hurricanes and storms which have not only provoked floods, but landslides in steep mountains, killing thousands of people, the most recent example being hurricane Stan in 2005, which buried a small rural town on the foothills of Toliman volcano, killing nearly two thousand people. In addition, Guatemala is one of the many countries of the Pacific Ring of Fire. The country lies at the intersection of three active tectonic plates: Caribbean, North American, and Pacific. Throughout the centuries, earthquakes have destroyed cities, and forced governments to translate the centers of government from one place to another. The recent 1976 earthquake killed over 25 000 inhabitants, in a 'text-book example' of structural vulnerability of adobe houses which collapsed burying entire families at once.

In 1969, as a consequence of the devastation provoked by hurricane Francelia, the government of Guatemala established the National Emergency Council with the mandate to respond to such disasters. Almost 20 years later and within the framework of the International Decade for Natural Disaster Reduction IDNDR, this agency was transformed into the National Coordinating Agency for Disaster Reduction CONRED to incorporate notions of risk management and disaster prevention. To tackle issues related to risk management within this agency, the author established the risk management division, and promoted risk assessment and early warning as key activities.

The need for risk assessment stems from the fact that in many developing countries resources are very scarce, and thus there is a need to set priorities regarding how to invest such resources to minimize existing and potential risks (VILLAGRÁN, 2001). The logical strategy is to invest such resources initially on high-risk zones, which implies having gone through an exercise of qualifying and quantifying risks. This last phrase, easier to write than to execute, is the main issue discussed in this paper.

2 Risk assessment

Within the context of natural disasters, risk assessment encompasses hazard and vulnerability assessments. Combining these two factors to generate a risk indicator seems to be even more difficult to handle. Nevertheless, regardless of the methods to measure such indicators, a basic end result must be satisfied when assessing risks:

- A community labeled under the category of high risk should experience a disaster when the hazard materializes as an event.
- A community labeled under the category of low risk should only experience minor damages when the hazard materializes as an event, tending to an emergency.

In this context, the difference between an emergency and a disaster is related to the need to request assistance from an external level. An emergency is handled without having to request external assistance of any kind, while a disaster will indeed require external assistance for its handling. In addition, it would be desirable to have a method to calculate such indicators that could also be employed in a theoretical, back-track fashion to display or explore specific measures which can be implemented to reduce existing risks. For example, the author designed a method to evaluate the structural, functional, economic income, and gender vulnerability of individual houses in terms of different parameters. In the case of structural vulnerability, it was assessed through its components: walls, roof, doors, windows, etc. (VILLAGRÁN, 2005). The characterization of the vulnerability was made initially ranking such components with respect to each other and assigning different degrees of susceptibility to damage (low, medium, and high) to different types of construction materials used to build such components. Numerical values were assigned to the components of the houses and construction materials according to historical records of disasters. More vulnerable materials were assigned larger values, while low-vulnerable materials were assigned low values. Functional vulnerability was assessed in a similar way, identifying the main components and then characterizing the susceptibility of each component via the different options commonly available and employed within towns and villages.

Once the vulnerabilities of different types had been assessed with the use of a survey specifically designed for this purpose, GPS measurements of the household's compound were carried out to generate a GIS layer related to vulnerability. To assess the risk, hazard maps were gathered from the literature and also framed in a GIS layer. As risk assessment involves the combination of vulnerabilities, hazards, GIS offers a possibility to make this combination in a geographical format. In the simplest case, risk is assessed via the direct combination of hazards and vulnerabilities, as presented in the following table (VILLAGRÁN, 2004):

Table 1: Table to calculate risks as a function of hazards and vulnerabilities

Vulnerability Hazard	Low	Medium	High
Low	Low risk	Low risk	Medium risk
Medium	Low risk	Medium risk	High risk
High	Medium risk	High risk	High risk

3 Risk assessment at the household level

Figure 1 displays results carried out with respect to the assessment of risks associated to landslides in an urban settlement in Guatemala City. This assessment was conducted during a pilot project sponsored by the Inter-American Development Bank and the World Bank, and was executed by the Division of Risk Management of CONRED in the period 2000–2002. It spanned 9 human settlements in four municipal districts encompassing the larger Guatemalan metropolitan district.

Geologists from the National Institute of Seismology, Vulcanology, Meteorology and Hydrology – INSIVUMEH – and of the Division of Risk Management of CONRED carried out the assessment of the landslide hazard. Factors to characterize the hazard included geological formation types and slope. The hazard associated to landslides was characterized in low (green), medium (yellow), and high (red) degrees. The top map in figure 1 represents the hazard.

Vulnerability assessment was carried out with the support of young volunteers of CONRED composed of a group of university students which support institutional activities and receive training on disaster management and response. The second map displays vulnerabilities of individual households. Each plot of land was surveyed with a GPS and a questionnaire. As in the case of hazards, vulnerabilities were classified in three levels.

Deficiencies in preparedness included the lack of a local emergency committee, lack of an emergency plan regarding disasters, and a lack of an early warning system. In contrast to vulnerabilities, such deficiencies in preparedness applied to the entire human settlement. As in previous cases, three levels were employed: low, medium, and high.

Hazards, vulnerabilities, and deficiencies in preparedness were combined using the ARC-View software within CONRED. As a result, individual households were categorized in terms of low, medium, and high risk. High risk could result from the location of the household in a high hazard area and because of its high degree of vulnerability. Being an urban settlement, most of its households could be classified as to be under medium or high risk, and very few were categorized as low risk areas.

The project concluded with the dissemination of the results in all nine human settlements, the implementation of some measures to enhance disaster preparedness, and the dissemination of information to municipal authorities, in particular the Municipal Coordinating Agencies for Disaster Reduction.

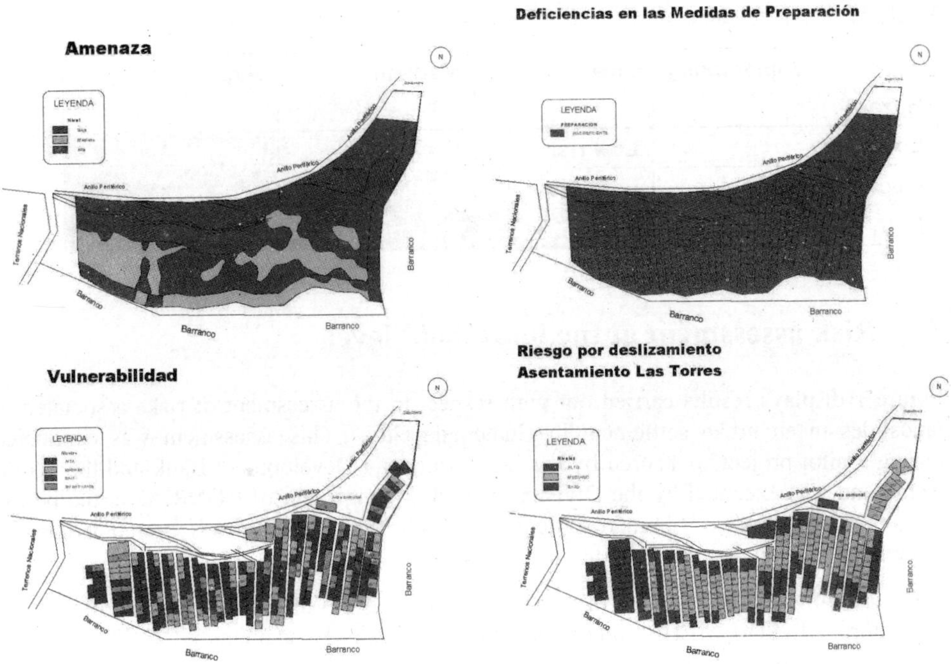

Fig. 1: Plots representing hazard, vulnerabilities, deficiencies in preparedness, and risk for the Las Torres urban settlement in Guatemala City. Source: MINER, 2002.

4 Risk assessment at the community level

In another pilot effort and at the request of the international non-government agency Action-Against-Hunger, the author was requested to carry out a multi-hazard risk assessment of communities within the municipal districts Camotan and San Juan Ermita belonging to the Department (Province) of Chiquimula in the eastern region of Guatemala, close to the border with Honduras (VILLAGRÁN, 2004). Using the same methodology described above, surveys were conducted for a sample of houses in more than 30 rural and urban communities of the two municipal districts in order to generate a database regarding vulnerabilities for communities with the capacity to express the percentage of houses categorized in low, medium, and high ranges.

However, a problem arises when trying to represent such type of information in a map using the poligons of the villages due to the fact that the non-uniform polygons cannot be split easily to represent these three proportions within the polygon. The only way in which this operation was accomplished was via the use of pie-charts linked to the communities. Thus, additional ways to map vulnerabilities and risks should be devised taking into account such needs. Figure 2 presents a map representing the structural vulnerability of communities with respect to earthquakes within these two municipal districts. The typical adobe houses, in a hazardous area lead to the medium and high risk conditions represented in the map.

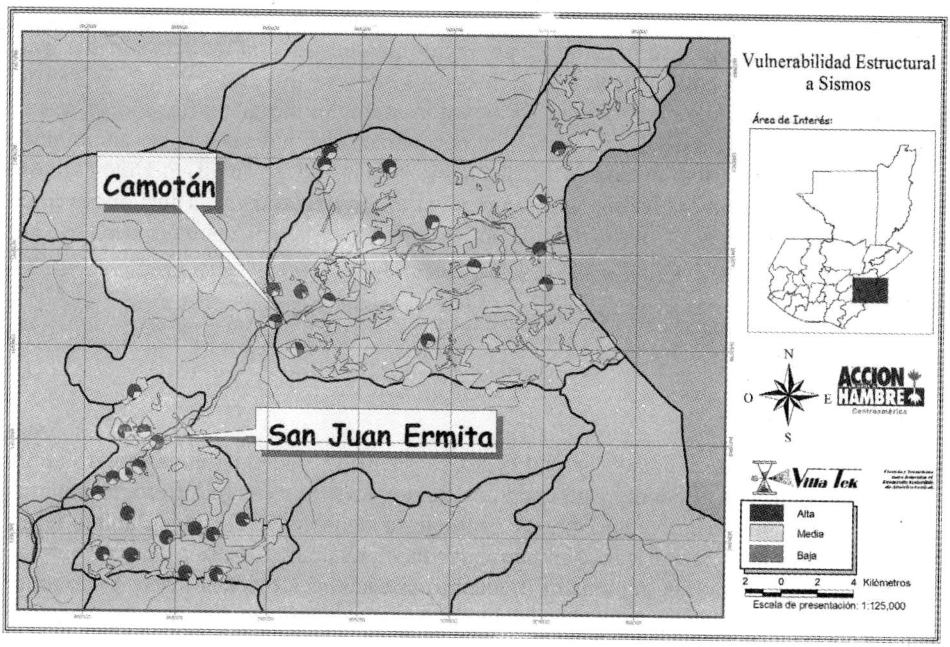

Fig. 2: Map displaying the proportion of households having low, medium, and high structural vulnerabilities with respect to earthquakes in the San Juan Ermita and Camotán municipal districts of Chiquimula, Guatemala. Source: VILLAGRAN DE LEON, 2004.

5 Normalized risk in the case of the housing sector

When generating vulnerability and risk information regarding the housing sector in communities of different sizes within a municipal district or within a province, an issue that emerges is the fact that larger communities can display large numerical vulnerabilities just on account of a greater number of houses. However, from the point of view of disaster management there is a need to classify communities in terms of possibilities of emergencies or disasters in relation to whether the proportion of houses at risk is either low or high within the community.

In an effort to deduce such information from standard risk maps, the author has defined the concept of *normalized risk* (VILLAGRÁN, 2002, 2004) which is obtained in dividing the total risk for the community by the number of houses within the community. In this sense, a community labeled with a high normalized risk would be expected to undergo a disaster should the hazard become a real event.

Figure 3 presents two maps displaying risk and normalized risk with respect to floods for communities situated in the Escuintla Department located in the southern part of

Guatemala. Floods are typical in this Department, as several rivers inundate the plains during hurricanes and tropical depressions. Because such an assessment spans already hundreds of communities, it is not feasible to use the household approach, but rather use the census data as baseline. Within countries like Guatemala, census at times comprises not only population, but also housing. Data regarding construction materials of the walls, floors, and roofs are collected during such census surveys. While one has to sacrifice precision, the use of such census data allows, for example, the quantification of risks in communities at the national level. While the normalized risk indicates those communities likely to undergo disasters in case of floods, some villages are so small, that they are only displayed as dots or tiny polygons when compared to the larger towns. Thus, one is forced to seek different scales of magnification to visualize such small communities, but here again, the use of GIS facilitates such a task, as visualization becomes easy using such software.

6 Discussion

Risk assessment in the context of natural disasters is carried to identify communities at high risk for two purposes: to identify geographical regions where to target efforts to minimize such risks through preventive measures to promote sustainable development; and to inform decision makers at different levels, in particular those related to disaster management and disaster response, regarding regions of municipal districts or provinces where disasters are likely to take place. In this case, humanitarian assistance will be required as it is to be expected that communities labeled in high risk will surely require such humanitarian assistance, as such communities are not able to cope with the expected damages and losses.

GIS tools are providing means to combine hazards, vulnerabilities, and other information provided that the base cartography has been digitized beforehand. And while GIS offers a unique capacity to display such information, recent pilot efforts within Guatemala point out the need to continue refining the capacities of tools. Two challenges, which emerge when using such tools, are:

- The need to be able to represent enhanced information regarding communities, in particular when such information can be decomposed into percentages, and one still wishes to link such information into the polygon related to the community.
- The need to visualize, within a single map, very small and very large communities to be able to compare levels of risk or normalized risk for targeting resources to communities labeled as high risk, regardless of their size.

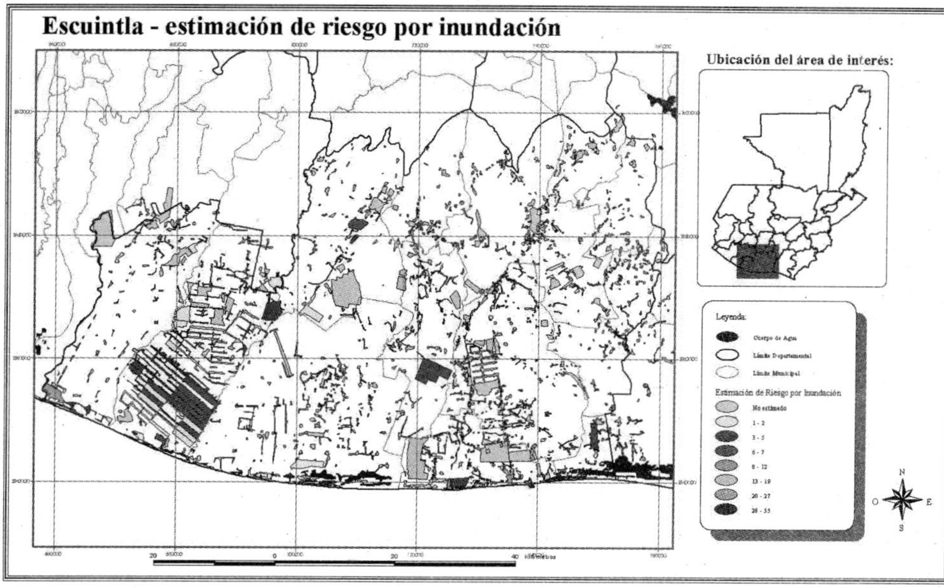

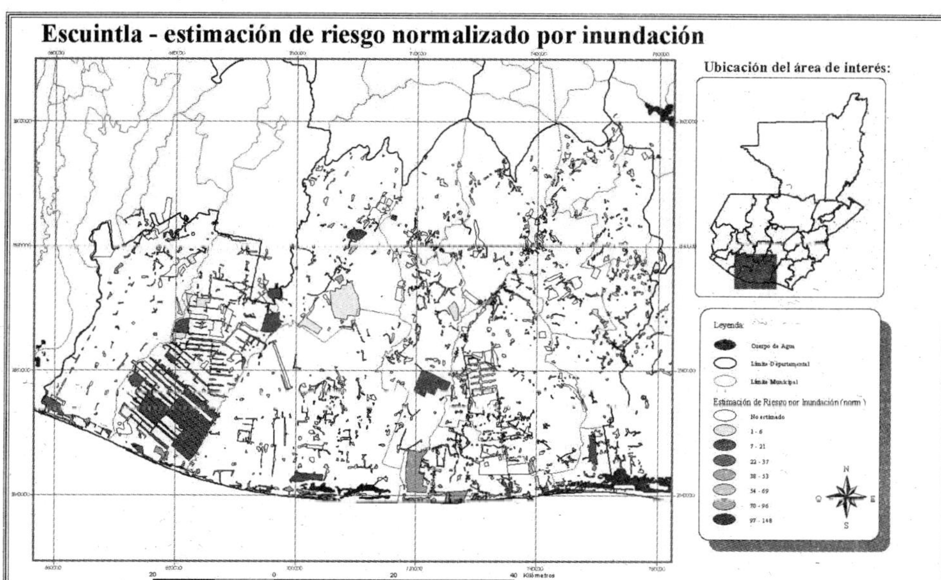

Fig. 3: Maps displaying risk and normalized risk with respect to floods in communities of the Escuintla Department in Guatemala. Source: VILLAGRAN DE LEON, 2002.

7 Conclusions

The need to develop indicators to evaluate vulnerabilities and risks is a 'must' for developing countries, which need a strategy to invest scarce funds to maximize the results from risk reduction. However, there are still many issues which need further analysis in terms of methodologies to evaluate and represention of risks and their components before they can be disseminated for use worldwide. Such issues should be the focus of research at UNU-EHS and other institutions.

One approach to promote the development of indicators could be focused on the systematization of the existing and new methods to calculate such indicators and testing their validity in the case of disasters via pilot zones. This would require the set-up of joint collaborations among the various groups around the world engaged in this line of work, so that methods simultaneously are shared and tested in various pilot zones to evaluate their precision in relation to forecasting disasters as social events.

References

MINER FUENTES, Y.S., 2002, Determinación de Riesgo: análisis y generación de mapas. Internal report prepared for CONRED (Coordinadora Nacional para la Reducción de Desastres), Guatemala.

VILLAGRÁN DE LEÓN, J.C., 2005, Quantitative Vulnerability and Risk Assessment in Communities in the Foothills of Pacaya Volcano in Guatemala. Journal of Human Security and Development, 1(1).

VILLAGRÁN DE LEÓN, J.C., 2004, Manual para la estimación de riesgos asociados a varias amenazas. Publication sponsored by Action Against Hunger. Guatemala.

VILLAGRÁN DE LEÓN, J.C., 2002, Reconocimiento preliminar de riesgos asociados a varias amenazas en poblados de Guatemala. Internal report prepared for SEGEPLAN (Secretaría General de Programación y Planificación de la Presidencia de la República de Guatemala), Guatemala.

VILLAGRÁN DE LEÓN, J.C., 2001, La naturaleza de los riesgos, un enfoque conceptual. CIMDEN (Centro de Investigación y Mitigación de Desastres Naturales), Guatemala.

Geoinformation for Poverty Alleviation – Panel Discussion in the Symposium 'Geoinformation for Development'

E. Alaphia WRIGHT

Summary

Development is multi-faceted – covering such aspects as agricultural development, industrial development, and information and communication technological development, to mention three evident examples. Effective collection, storage, processing, and dissemination of geoinformation is advancing at an impressive rate. The usage of geoinformation in catalyzing development is a challenge. A useful management tool that can be employed to tackle this challenge in the design stages of any geoinformation project or programme is the Results Based Management (RBM) Programme Logical ScoreCard. The discussion illustrates the use of the RBM Programme Logical ScoreCard in addressing the challenge of usage at the design stage for geoinformation programmes aiming at poverty alleviation.

1 Introduction

First of all, let me congratulate you for your choice of such an important theme for the symposium. Development in the broadest sense is by far one of the most challenging issues facing the world in the early part of this new millennium. Development is multi-faceted – covering such aspects as agricultural development, industrial development, and information and communication technological development, to mention three evident examples. You will notice that the three illustrations here can be viewed as reflecting a progression; from the agricultural development era, through the industrial revolution, and now into the Information –Technology – Communications (ITC) -age. The examples can also be viewed from the perspective of co-existence in that agriculture did not cease because of the emergence of manufacturing industries, and certainly industries have not all shut down because of the progress made in ITC. The three co-exist and they mutually influence each other.

2 agitSPECIAL

The various presentations here today actually reflect the wide range of possible interactions of the said co-existence. Effective collection, storage, processing, and dissemination of geoinformation is seldom contemplated without suitable information and communication technologies. This is evident from an examination of the topics presented here today, for instance:

- The mapping of water potential … to support refugee operations in Darfur,
- Spatial Decision support for Tropical Agriculture and
- Expert System for the Operative Environmental Diagnostics.

3 The challenge of usage

However, the most challenging aspect with all these is that of the usage of geoinformation in actually catalyzing development. Very often geoinformation projects and programmes do not give serious thoughts to the usage of the information they assemble or generate, and as such opportunities for contributing to development are lost. Fortunately there are simple programme design tools that can be used to cross-check whether issues of the usage of the geoinformation concerned have been taken into account in the design of given geoinformation programmes and projects. One such tool is the Results Based Management (RBM) Programme Logical ScoreCard (Fig. 1).

The RBM Logical Scorecard The minimum set of Evaluation Questions to assess the theory or logic behind the performance of a system such as an Organization, a Department, a Programme and a Project practicing Results Based Management (RBM)	
5	• Who 'enjoys' the ultimate benefits in the cases in question (with what arrangements)? • What are the ultimate 'benefits' these cases CONTRIBUTE to?
4	• Who uses the outputs (with what arrangements)? • What are 'they' to do 'differently' (with the help/use etc. of the outputs)?
3	• What are the outputs? • Who participates in delivery of the outputs (with what arrangements)?
2	• What are the processes? • Who participates in the processes (with what arrangements)?
1	• What are the inputs? • Who provides the inputs (with what arrangements)?

Fig. 1: The Logical Scorecard and the Ten Evaluation Questions for assessing the theory or logic underpinning the performance of a programme

4 The RBM Logical ScoreCard

The RBM Logical ScoreCard (shown in Fig. 1) presents a checklist of ten questions to be posed and answered in the design of a programme or project. Briefly, the RBM Logical ScoreCard gives the minimum set of evaluation questions to assess the theory or logic underpinning the performance of a system such as an organization, a department, a programme, or a project practicing (or implemented on the basis of) Results Based Management (RBM). The ScoreCard is normally read from the bottom to the top, beginning in row number 1. The attentive reader would have noticed that the five 'levels' captured by the RBM Programme Logical ScoreCard correspond to the five levels of a programme results-chain, namely: (1) inputs – then – (2) processes – then – (3) outputs (or immediate results) – then – (4) outcomes (or intermediate results) – then – (5) impacts (or end results).

5 Illustration, poverty alleviation

Let me now illustrate the use of the RBM Logical ScoreCard in the context of a geoinformation programme or project aiming to influence poverty alleviation. Our starting point is that the given geoinformation programme actually delivers 'improved geoinformation for poverty alleviation' as the main output. Without delving into the mechanics of actually producing the improved geoinformation, figure 2 shows an extract of the RBM Programme Logical ScoreCard in which the inputs (level-1) and processes (level-2) have been omitted. We thus start at the output (immediate results) level-3. The second question in the output level is straightforward enough: 'who participates in the delivery of the outputs?' Answer: 'The Geoinformation R&D team from Institute XXXX'. We start encountering problems when we move to level-4, and ask: 'Who uses the outputs?' Invariably there are no definite answers. Sometimes the answer will bc: 'other researchers', or quite broadly 'policy-makers'. Now, this is generally unsatisfactory. When one does not really know who should use the outputs, how can one know what the users are supposed to do differently? The issue gets even more complicated when we get to level-5: 'who enjoys the ultimate benefits of the cases in question?' and 'what are the ultimate benefits to which the cases contributc?' These questions have been answered with a series of question marks in the example given here (Fig. 2)

The RBM Logical Scorecard Example: Geoinformation and poverty alleviation (Extract)		
	Questions	**Answers**
5	• Who 'enjoys' the ultimate benefits in the cases in question? • What are the ultimate 'benefits' these cases CONTRIBUTE to?	• ??? • ???
4	• Who uses the outputs? • What are 'they' to do 'differently'?	• ??? • ???
3	• What are the outputs? • Who participates in delivery of the outputs?	• Improved geoinformation for poverty alleviation • Geoinformation R&D team from Institute XXXX

Fig. 2: Example of The RBM Logical Scorecard for a geoinformation programme aiming to influence poverty alleviation

6 Discussions

The fundamental issue here is that the results chains of the said geoinformation programme are incomplete, and as such it becomes almost impossible to determine whether the said project will actually contribute to poverty alleviation. Fortunately, with regards to poverty alleviation the set of issues to be tackled at both levels 4 and 5 are consistently stable. For instance, when one asks the question: 'who uses the outputs?' The answer has to be: 'the people of a given locality, region, country etc.' and, the answers to the question: 'what are they to do differently?' would be: 'they are to be constructively engaged in, and benefit from their activities in the areas of:

- Income generation,
- Food security,
- Health,
- Management of environmental resources,
- Education, culture and leisure,
- Business/enterprise and
- Governance.

Overall a given programme can contribute to poverty alleviation if and only if its outputs can be instrumental in supporting the actions of various groups in actually performing well in at least one of the seven areas above. If the outputs are not going to be instrumental in supporting development in any of these seven areas, then the programme or project would have to be redesigned if it is expected to contribute to poverty alleviation.

A related issue is that of baseline data with respect to the expected usage. Even when the question as to who are to use the outputs have been answered, it is important to be clear about the usage or non-usage at the start, namely the baseline situation.

Finally the ultimate benefits from constructive engagement in the said seven areas in all cases would be dignity, peace, and prosperity! Hence a geoinformation programme or project aiming to contribute to poverty alleviation should in effect be able to contribute to dignity, peace, and prosperity – the ultimate goal in development.

7 Concluding remarks

Geoinformation is indeed important for development – development being understood in its widest sense. In terms of poverty alleviation however, the usage of improved geoinformation presents an interesting challenge for geoinformation projects and programmes. Fortunately, this challenge can be tackled by systematic consideration of questions regarding expected usage of the said geoinformation at the design stage of given geoinformation projects and programmes. The RBM Logical ScoreCard is a simple evaluation tool useful for examining the logic of given programmes with respect to the completeness of the results-chains concerned. Gaps identified in the results-chains can be corrected, thus enhancing the prospects of successful projects and programmes. Mr. Chairman, colleagues in the panel, ladies and gentlemen, I thank you for your attention.

References

WRIGHT, E.A., 2002, Re-engineering of interlocking log frames – an irrigation project design case. Physics and Chemistry of the Earth, 27, pp. 1005–1010.

WRIGHT, E.A. and WRIGHT, B.S., 2002, Practical Strategic Planning and Policy Development using Systems-ware Log Frames. Short Course, Harare, pp. 400.

WRIGHT, E.A., 2003, The re-design of an integrated water and pollution management programme using the systems-ware model of the log frame. Physics and Chemistry of the Earth, 28, pp. 973–984.

WRIGHT, E.A., 2004, Re-thinking Development Evaluation – don't neglect the fundamentals, 3rd African Evaluation Association (AfrEA) Conference, Cape Town, November 2004, 8 pp.

Geospatial Information, Sustainable Development and Decision Making in Developing Countries

David STEVENS

Summary

As the environmental economist, Kenneth E. BOULDING, once stated (1974): 'The world moves into the future as a result of decisions, not as a result of plans'. We can produce better plans, more detailed understanding of the possible consequences of our actions, but unless the correct choices are picked and implemented, all the planning done will result in nothing more than additional paper to be recycled. We can use geospatial technologies to support 'better sustainable development planning', but unless the final decisions reached and implemented actually do lead to an improvement of the well-being of the population affected by the outcome, using such technology solutions will just needlessly increase the cost of the planning activities. This paper reflects on this and sets out to suggest guidelines to deal with the concern most geospatial-technology practitioners have, which is how we can maximise the positive impact that geospatial information could have on the sustainable development process in developing countries.

1 Geospatial information

There has been an exponential increase in the availability of geospatial information. Today we do not think twice about the weather map shown every night on the evening news, which is made possible because of the real-time information the numerous meteorological satellites now provide us with. Nowadays anyone can 'hit the trails' with a handheld GPS that includes contours, elevations, detailed terrain features, major lakes and waterways and extensive points of interest, that makes navigating the roads and trails easier than ever and helps find sights and services along the way.

But all this geographic information and the tools to process the data are relatively new. Available today are technologies that did not exist just a generation ago such as data gathered from space-based sensors, global navigation satellite systems, communication satellites and meteorological satellites. And in order to integrate all these data specific information technology solutions were developed. Not only software solutions such as geographic information systems (GIS) and image processing software but also hardware solutions such as plotters, scanners, and hand-held devices which enable us to use these cutting edge solutions anywhere in the world.

Moreover, not only those living in developed countries can access and use such cutting-edge solutions but virtually anyone in the world. Images from meteorological satellites are now directly accessible by National Meteorological Organisations and in fact by any interested institution. Satellite imagery obtained from a number of space-based sensors such as Landsat and MODIS can be freely downloaded (or at a minimum cost) by anyone that has access to the Internet. Even those that have a slow Internet connection can receive such data on CDs, DVDs or external hard drives. There are free GIS software packages downloadable from the web such as SPRING (www.dpi.inpe.br/spring/english/) and GRASS GIS (http://grass.itc.it/) or at a minimum cost such as IDRISI (www.clarklabs.org/).

As will be discussed further on, more information improves the planning process, contributing to better decision-making. But despite this wide-availability and easy access developing countries still lag significantly behind more advanced nations in the local adoption and application of such technologies (GEORGE, 2000).

2 The impact on the sustainable development process

So many of the problems we observe today in developing countries are the result of decisions made (or not made) in the past, the same way the near future will bring additional critical problems which will be the results of the decisions being made (or not made) today. Geospatial technologies in general and Geographic Information Systems in particular have been considered valuable tools for supporting the sustainable development process, but not enough consideration has been given to what exactly the technologies actually have to offer to influence the outcome of the final choices made and the definite implementation of these choices. In order to incorporate successfully these technology-based planning support solutions, there is a need to focus on the actual decision outcomes and not on how the planning component of the search for 'better' solutions can be improved through the incorporation of such technologies.

The components of this social process termed decision-making are data, decision models, the decision environment and people; more data leads to more information, and decision models go a step further by transforming information into knowledge of consequences. By understanding how the rational actor uses geospatial technologies to transform information into knowledge and the changes in the decision environment due to the introduction of the technology one would be able to improve initial design, which in turn will lead to better decision-making, the ultimate objective for having the technology in the first place.

The initial thirty years of research in GIS technology focused on the establishment of technical solutions for managing spatial data, constrained by the available hardware, then turning toward the need to increase analytical functionality. Only recently has the GIS community realized that its knowledge of how decisions actually do improve with the technology is in fact limited. Research has moved from the technical to the scientific (consolidating GIS as science) and presently to the social, with the aim now being to try to maximize the social benefits.

One of the social benefits is 'better' decision making, which since the beginning has always been considered 'the promised land' (MAYNARD, 1992) of GIS, but not much research has been done to try to understand how (and even if) the introduction of GIS technology impacts the decision process.

2.1 Impact on the decision process

Geospatial technologies facilitate better decisions by providing access to more information and by rationalizing the decision process thus contributing to better decision-making. Better decision-making occurs when we do in fact implement a more satisfactory alternative that is coherent with the identified objectives of the spatial problem being tackled, narrowing the gap between the stated objectives and the solution chosen. These specific objectives, which are consciously or unconsciously backed by our moral and cultural environment, guide the range of possible alternatives and the evaluation of each one.

Research I carried out in Bolivia involving a group of end-users using a GIS-based decision support system provided an insight into the benefits that such a solution brings to the decision making process. Initial conclusions show that:

- using a GIS actually helps to clarify the decision process (providing structure to an unstructured decision process);
- a GIS enables the consideration for a broader range of alternatives and the possibility of addressing new questions;
- in the search for coherence between objective and chosen alternative there is a change of stated objectives, narrowing the gap between what is wanted and what is being proposed, which translates into 'better' decision-making;
- a GIS brings in the possibility of discussion and modification of the proposed decision criteria;
- a GIS enables the exploring of conflicting decision criterion and conflicting objectives, and the incorporation of methods for resolving these spatial conflicts.

2.2 Further thoughts on the social perspective of geospatial technologies

CHRISMAN (1996) in his contribution to NCGIA's Initiative 19 (GIS and Society: The Social Implications of How People, Space, and Environment are represented in GIS) argued correctly that in order to understand the linkages between GIS and society we must accept the three following principles:

'1. *Social context influences GIS, and GIS influences society.*

2. *Multiple social structures interact in this process.*

3. *GIS technology cannot be reduced to some mechanistically determined parts, the people act as agents.*'

Indeed, decision-makers responsible for finding and implementing better decision-making can be seen as agents acting rationally in their search for optimal decisions, but eternally bounded by social structures. The outcome of the decision process is a balance between how agents interpret (or choose to interpret) the output of the various decision models and the social structures which guide which models should be implemented in the first place.

2.3 Geospatial technologies must become an indispensable component

There is evidence that geospatial information is able to contribute to sustainable development by strengthening the planning process, which in turn contributes to improved decision-making. Because geospatial technologies are still not so widely used in developing countries, the benefit of the increasing availability of geospatial information is still not being felt. There is the notion that planners in developing countries do not know how to take advantage of geo-spatial technologies, or any other automated information system; the truth is though, the stretch has to be toward the planners (end-users) which detain the understanding of the spatial concepts and models that support the decisions that are being made today. If we cannot incorporate this decision process into a geo-spatial technology solution, we cannot expect potential planners to have any use for the technology.

During the process of defining these solutions many factors have to be considered including the principle that social structures influence how such technologies are used and vice-versa.

The most significant impacts of technology tend to occur when it becomes indistinguishable from the actions of everyday life. The technologies themselves become commonplace and are unconsciously accepted as somehow indispensable. As stated by VEREGIN (1995) 'The most significant impacts of the technology therefore tend to occur when the technology has ceased to be a 'tool' and has become an integral and indispensable component of everyday life'. Geospatial technology practitioners should strive to understand the social constraints and ensure that the solutions developed become an integral part of the decision-making process.

3 Understanding the limitations and constraints

At the same time, it is important that geospatial technology practitioners step back from time to time and remind themselves that there are limitations to the solutions they are proposing. As LAURINI and THOMPSON (1992) stated, spatial information systems represent a perspective, a way of looking at problems about the world and because of this singular perspective one must always have in mind that 'no matter how carefully packaged, no tool is perfectly neutral' (CHRISMAN, 1999). Building a geospatial database is more than just transforming relevant data into a digital format. The compilation process of the geospatial database can act as a filter or a constraint; the resulting database invariably is selective data – necessarily incomplete and generalised. Sometimes this is due to technological constraints, which dictates what can be included, and sometimes it stems from the social context, which selects the data that should be included and omitted.

Additionally there is the need to understand (1) the dimensions of error in the data: geometric and attribute. These errors appear when inputting the data into a GIS environment, and gain importance in such an environment. (2) The adequacy of the models for the stated objectives – it's so tempting to run a model just because it comes with the software or it can be easily developed, but the information provided by the model might not be adequate for evaluating alternatives. (3) the need to consider the uniqueness of places – let's not repeat the same errors of the quantitative revolution in geography. And finally, (4) the ability one needs to be able to understand the products obtained from a GIS, not everyone can 'read' a map – this leads to the need to improve visualization and display tools, and user interfaces.

Perhaps the most significant concern is that the use of geospatial technology solutions can mislead the end users. Information can be distorted by how it is represented in the database and by what is included or excluded. The real danger about GIS data is that people may come to believe that everything they see is truth.

Additionally, there is an existing notion that just because a solution has an IT component then it will be successful. This misconception goes hand in hand with the thought that 'if it is digital then it is correct'. MARKUS (1997) discussed the issue of technological determinism arguing that most implementers of IT solutions (the managers and the IT specialists) 'seem to believe in magic'. These adopters describe IT as a magic bullet and they believe that by initiating or developing IT solutions they are building the gun that will fire this magic bullet. He correctly points out that the flaw in this theory is that there is no definition on who is actually aiming and firing the gun. 'It doesn't say who is going to be the change agent responsible for producing the desired outcome with the magic bullet'.

Those who believe in the magic of IT solutions assume that the gun fires itself, not really understanding that the theory hides one of the most important characteristics of information technology and that is that IT is a package of ideas of how people should work differently. Thus, the change comes from people and not the technology itself. In addition, change is made possible when it involves changing people's minds. People have to create change and IT contributes toward this because it delivers the one need for human empowerment – valid information.

4 Bridging the gap

Building upon these previous considerations that demonstrate the vital role geospatial information could have in the sustainable development process and the recognition that such a role is still not being fulfilled in developing countries, the following guidelines are being put forward as starting points to support the development of strategies that could contribute to the successful use of geospatial technology solutions in developing countries.

In the first years of GIS the term 'data bottleneck' used to be synonymous with the technology itself. Nowadays the shift has been towards a 'flood of data'. But developing countries still do not have access to this 'flood of data' and because of this a systematic compilation of relevant geospatial data and information or information on where such data and information can be obtained should be put together and made especially available to developing countries. This should coincide with the implementation of national spatial data infrastructures and data mining and information extracting tools, specifically developed for the individual needs of each end user.

Capacity building and strengthening of institutional arrangements at all levels is key to increasing the ability of organisations and individuals to effectively use geospatial technologies for sustainable development. Because of this, any strategy should include support to capacity building as a key cornerstone of the strategy: the use of new technologies is only possible if experts are provided with the needed training.

Focus on knowledge management and knowledge transfer is also central to a successful strategy. Knowledge management is concerned with the representation and processing of knowledge by humans, machines, organisations and societies. The aim of knowledge management is to ensure that the right knowledge is available in the right forms to the right end user at the time it is needed and at a suitable cost. Knowledge Management is unique because of its greater focus on the management of specific knowledge assets and development and cultivation of the channels through which knowledge flows (transfers).

This should be one of the main concerns of the geospatial technology practitioner: Understand how to increase the flow of existing knowledge to end users that should be accessing and using such knowledge in the development of sustainable development studies and support actions that facilitate this knowledge sharing. This includes the consolidation of a knowledge-base by identifying and bringing together knowledge, practical know-how, expertise and best practices, and by focusing on capturing and making such knowledge available to all end-users.

Finally, a successful strategy should recognise existing networks of excellence and strive to ensure that end-users participate in these networks. Examples of such networks include the communities of practice being implemented as part of the Global Earth Observation System of Systems (GEOSS) and the network of partners to be coordinated by the planned United Nations Platform for Space-based Information for Disaster Management and Emergency Response (SPIDER)

References

BOULDING, K.E., 1974, Reflections on Planning: The Value of Uncertainty. Technology Review, 77(1), p. 8.

CHRISMAN, N., 1996, GIS as Social Practice. Contribution to NCGIA I-19 GIS and Society: The Social Implications of How People, Space, and Environment Are Represented in GIS.

CHRISMAN, N.R., 1999, What Does 'GIS' Mean?. Transactions in GIS 3(2): 175–186.

GEORGE, H. 2000: Developing Countries and Remote Sensing: How Intergovernmental Factors Impede Progress. Space Policy, 16, pp. 267–273.

MARKUS, M.L. and BENJAMIN, R.L., 1997, The Magic Bullet Theory in IT-Enabled Transformation. Sloan Management Review, 38(2), pp. 55–68.

MAYNARD, G., 1992, From the Editor – The Promised Land. Geo Info Systems, 2(10), pp. 329–342.

LAURINI, R. and THOMPSON, D., 1992, Fundamentals of Spatial Information Systems. London: Academic Press.

VEREGIN, H., 1995, Computer Innovation and Adoption of Geography. In PICKLES, J. (Ed.), Ground Truth. The Social Implications of Geographic Information Systems. The Guilford Press, New York.

Towards Integrated Disaster Risk Assessment for South Africa

Dusan SAKULSKI

Summary

It was a long and tortuous road for the South African disaster risk legislation. From the 1998's Green Paper till 2002's Disaster Management Act, and 2005's National Disaster Management Framework. One of the basic principles of that legislation is to enable a 180-degree shift, from a passive (reactive) to the proactive approach, maximising prevention. One of the first serious implementation of the legislation is an integrated disaster risk assessment. It will include spatial and temporal multi-hazard, vulnerability assessment and risk analysis, for the municipal, district, provincial and national level. All South Africans should be aware of the risks they live with.

1 Introduction

Prior to the year 2000, disaster risk reduction in South Africa was, in its real nature, reactive and exercised as an emergency management. Legally, it was based on the old Civil Protection Act from the mid seventies. In 1994 the combination of natural hazard and unplanned and unsustainable development created misery in the country. That disaster became a catalyst for change in Government's attitude and initiated an inclusive, consultative reform process, resulting in the Green Paper on Disaster Management, published in 1998. The Government's commitment to this reform was evidenced in 1997 by Cabinet approval of the establishment of an Inter-ministerial Committee (IMC) for disaster risk management, consisting of Cabinet's ministers carrying portfolios relevant to the disaster risk management. The committee was mandated to table the Green Paper on Disaster Management for Cabinet approval. The Green paper was followed by the White Paper on Disaster Management, which was gazetted in 1999, and the Disaster Management Bill, tabled in Parliament in the Year 2000. The Disaster Management Act (Act 57 of 2002) was promulgated and, finally, the National Disaster Management (Policy) Framework, which was to guide the implementation of the Act, was gazetted in 2005.

2 National Disaster Management Framework

The National Disaster Management Framework comprises four key performance areas (KPAs) and three supportive enablers required to achieve the objectives set out in the KPAs. The KPAs and enablers are defined by specified objectives and, as required by the Act, key performance indicators (KPIs) to guide and monitor progress. In addition, each

KPA and enabler concludes with a list of guidelines that will be disseminated by the NDMC to support the implementation of the framework in all three spheres of government.

Key performance area 1 focuses on establishing the necessary institutional arrangements for implementing disaster risk management within the national, provincial and municipal spheres of government. It specifically addresses the application of the principle of cooperative governance for the purpose of disaster risk management. It also emphasises the involvement of all stakeholders in strengthening the capabilities of national, provincial and municipal organs of state to reduce the likelihood and severity of disasters.

Key performance area 2 addresses the need for disaster risk assessment and monitoring with the aim to set priorities, guide risk reduction action and monitor the effectiveness of our efforts. Although South Africa faces many different types of risk, disaster risk specifically refers to the likelihood of harm or loss due to the action of hazards or other external threats on vulnerable structures, services, areas, communities and households.

Key performance area 3 introduces disaster risk management planning and implementation to facilitate development-oriented approaches, plans, programmes and projects in striving to reduce disaster risks. KPA3 addresses requirements for the alignment of disaster management frameworks and planning within all spheres of government. It also gives particular attention to the planning for and integration of the core risk reduction principles of prevention and mitigation into ongoing programmes and initiatives.

Key performance area 4 presents implementing priorities concerned with disaster response, recovery and rehabilitation. KPA4 addresses requirements in the Act for an integrated and coordinated policy that focuses on rapid and effective response to disasters and post- disaster recovery. When a significant event or disaster occurs or is threatening to occur, it is imperative that there must be no confusion as to roles and responsibilities and the necessary procedures to be followed.

Enabler 1 focuses on priorities related to the establishment of an integrated and comprehensive information management and communication system for disaster risk management. More specifically, it addresses the information and communication requirements of each KPA and Enablers 2 and 3 and emphasises the need to establish integrated communication links with all disaster risk management players in national, provincial and municipal spheres of government.

Enabler 2 addresses disaster risk management priorities in education, training, public awareness and research. This enabler describes mechanisms for the development of education and training programmes for disaster risk management and associated topics. It addresses requirements to promote and support a broad-based culture of risk avoidance through strengthened public awareness and responsibility. It also discusses priorities and mechanisms for supporting and developing a coherent and collaborative disaster risk research agenda.

Enabler 3 sets out the mechanisms for the funding of disaster risk management in South Africa.

3 Disaster Risk Assessment and Risk Reduction Planning

Disaster risk assessment (KPA 2) is the first step in planning an effective disaster risk reduction programme. It examines the likelihood and outcomes of expected disaster events. This would include investigating related hazards and conditions of vulnerability that increase the chances of loss. Disaster risk assessments must be undertaken to anticipate and plan for known hazards or disasters to prevent losses and limit endangering impacts, as well as to ensure that development initiatives maximise their vulnerability reduction outcomes.

For the past few decades, activities related to disaster risk reduction research and implementation were undertaken in the area of hazards (type, frequency, magnitude, return period, ...). The most frequent ones are water related hazards, such as floods ('National Hazard and Vulnerability ATLAS'; SAKULSKI, 2002, Figure 1), droughts and water related diseases (i.e. cholera).

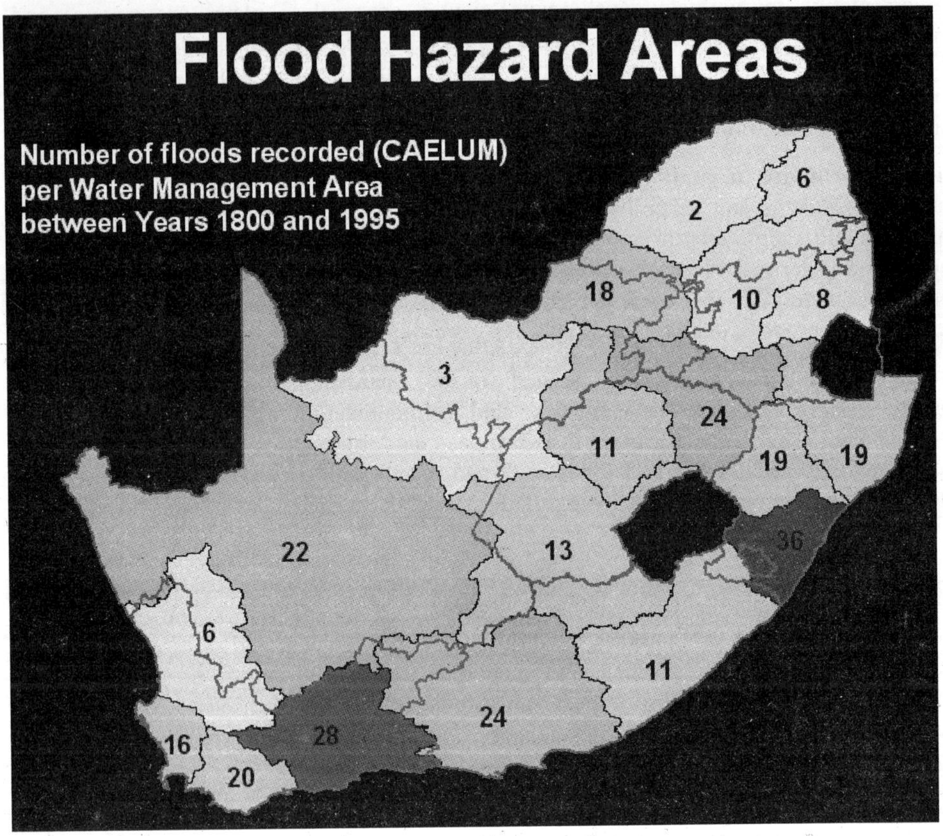

Fig. 1: Flood hazard areas for South Africa (SAKULSKI, 2002)

The vulnerability component is very new amongst disaster risk reduction researchers and practitioners, in South Africa. They did refer to vulnerability in the past, but it was more due to terminology confusion rather than real vulnerability research.

According to the Framework, here are some of the situations that require disaster risk assessment, focusing on vulnerability:

Table 1: Key vulnerability criteria

Key vulnerability criteria to achieve	Examples of where disaster risk assessments must be done
Increased sustainability of a development project or programme to support vulnerable households	As part of the planning for an infrastructural development, for example, assessing the likelihood of weather, flooding, subsidence and other threats damaging the structure, so that these can be factored into the construction specifications.
Reduction of potential harmful consequences associated with industrial, commercial or other delopents.	As part of environmental impact assessments for large-scale developments, including industrial, commercial and other enterprises that may increase disaster risk.
Increased understanding of a rapidly changing risk for imroved disaster risk management planing.	In a flood-prone estuarine area that has recently experienced considerable population growth and is facing increased coastal erosion.
Increased robustness of developent initiatives in poor communties and areas.	In an informal settlement characterised by recurrent small and medium-size disaster losses that undermine assets and livelihoods.
Management of high-risk periods and conditions to ensure service and/or business continuity.	Electricity transmission lines and rail infrastructure, as well as health and emergency services, to ensure these essential services do not 'fail' under expected high-risk conditions.
Provision of appropriate support for at-risk activities, services, areas, communities and houseolds following an 'alert'.	Following a drought warning or cholera alert in rural areas, to identify communities and households most at risk and to focus or target preparedness and response actions.

A reliable disaster risk assessment for a specific threat should be able to answer the following questions:

- How frequently can one expect an incident or a disaster to happen?
- Which areas, communities or households are most at risk?
- What are the likely impacts?
- What are the vulnerability or environmental and socio-economic risk factors that increase the severity of the threat?
- What capabilities or resources exist to manage the risk?
- Is the risk becoming more serious?

- Is the risk undermining development progress in the areas, communities and households it affects?
- If so, is the management of the risk a development priority?
- In the areas and communities affected by the risk, are there any other significant risks?

References

DISASTER MANAGEMENT ACT (Act 57 of 2002), Republic of South Africa.

NATIONAL DISASTER MANAGEMENT (POLICY) FRAMEWORK, 2005, Republic of South Africa.

SAKULSKI, D., 2002, National Hazard and Vulnerability 'ATLAS'. (sandmc.pwv.gov.za/atlas/)

Existing Infrastructure and Available Resources in Support of Remote Sensing Deployment Within Africa

Tsehaie WOLDAI

Summary

Different Nations in Africa, stratified by their degree of development, have different points of view of how to use and manage their resources. In its broadest term, the different Information Communities bring a different meaning to remote sensing (RS) technology, and for that reason, have different approaches of producing, accessing and managing Geoinformation derived from this technology at the national level. They operate on a stand-alone, extremely sectoral basis where they lack inter – as well as intra-organizational links.

A National Geographic Information Community is composed of organizations and loosely connected users, e.g., cartography community, cadastral, geology, agriculture, natural resources, etc. all working in isolation, poorly linked and duplicating their efforts. With jurisdiction sentiments and data secrecy culture prevailing in most African countries, it is hardly possibly to consider the existing systems and approaches adopted by the various communities as reliable sources that can channel information to the regional planning domain. This situation put a great impediment on the nations' effort to tackle development problems in a systematic and reliable spatial information foundation.

Efficiency of the planning system needs accessible, affordable, adequate, accurate and timely spatial and non-spatial information. Information sharing in turn needs an efficient route that can provide access to the needy. The potential route can be achieved and accessed through the implementation of a well organized and structured remote sensing infrastructure, as a remedy and wheel towards good information management and governance. 30 years after the introduction of remote sensing technology in sub-Saharan Africa, however, existing infrastructure and available resources in support of remote sensing on the continent remain much to be dreamed off. The root-cause for such problems lies in the way the technology was introduced into the continent and the lack of dedicated government support to the technology itself. Here below, the African remote sensing legacy is first introduced so as to appreciate and understand the existing infrastructure and available resources in support of remote sensing development within Africa.

1 The African Remote Sensing (RS) Legacy

The pre-1980 approaches towards remote sensing technology in Africa can be summarized as (a) supply- and (b) technology-driven with an 'put the data in the computer, we will figure out later what to do with it' and 'stop updating the system, the project is over' attitude (WOLDAI, 2002a). This approach rarely met the needs of the end users. It also fell short from reaching the decision making process due to poor linkage with national institutions.

- There was a clear imbalance between the information produced and the national capacity to absorb and use it.
- Processing & interpretation was done outside Africa and only the final product was delivered in analogue form.
- Training and education of African end-users received lip-service and remotely sensed data including software and hardware was rather remote and not made available for follow-ups.
- Most of the fieldwork data was never delivered and can hardly be traced. In countries like Kenya, Tanzania, Uganda, Ethiopia, Sudan for example as much as 90% (WOLDAI, 1995) of the information collected has no geospatial reference.
- The datasets collected then does not conform to any set of quality standards. This implies that the datasets that are useful if combined cannot be reproduced and its integration are at best difficult, and at worst impossible.

From 1980 to 1990, proponents of the technology in the First World advocated for: demand-driven, application-oriented and problem-solving approaches. Unfortunately, the data availability did not translate into greater use and greater use of data did not necessarily improve environmental sustainability. The demand driven approach was again a piecemeal approach – the categories of users were not well identified, nor have their specific needs been well surveyed (WOLDAI, 2002a). The technology was introduced through: map making and producing agencies, international or regional workshops and seminars, advancement of commercial companies dealing with satellite data and environmental, or remote sensing courses offered by institutions (colleges, universities, etc.) in Europe and North America. In this period, many problem- and application-oriented studies were undertaken.

- RS (maps) in analogue mosaic format however, were the model of supply of geosciences data and information.
- Field photographs, interpretation results, laboratory analysis, field notes, field data, unpublished maps and reports were poorly documented.
- Inconsistency, duplication and standardization problems persisted. Most analogue RS products were poorly georeferenced.

From 1990 to 1995, the introduction and expansion of GIS in different parts of Africa and the use of this technology by various organizations can be considered as encouraging. Organizations once confined to analogue map-making using hard-copy RS, slowly migrated to digital map delivery. World-wide improvement in software, increased storage capacity and low hardware costs enabled GIS and associated technology on desktops in many parts of the continent. When critically evaluated however, its contribution to support development efforts in a coordinated way is still questionable. Nor is the amount and type of geo-data that is made available to the users significant. This helps one at least to understand that the system is tied up with a number of critical problems that have hampered the possible application of RS in support of spatial decision making at various levels. GIS application in many of the African countries, was at its infancy stage.

- RS, by most was used in the search for land use, agriculture, mineral resources, for scientific research, for land-management decision making, engineering design, for assessing geological disasters, for groundwater exploration, etc. The surveying methods however, remain largely mono-disciplinary and focused on map production.

- Despite the complex task of acquiring data from the Earth surface using remotely sensed data supported by human observation in the field, only a fraction of the knowledge acquired in the latter are represented.
- Despite years of effort that went into making maps, only one interpretation was published. This data usually were and are still only structured in the mental archives and models of the surveyor or investigators and therefore lost forever when he or she retires from the organization (WOLDAI and SCHETSELAAR, 2002).

The years 1995 to present are crucial to the development of RS applications in Africa. Several technological innovations contributed to this endeavour:

- The advancement in Information Systems/Information Technology (IS/IT).
- Numerous geographic data were made available that are attributable to the growing of a number of related public, private institutions and centers in most parts of the continent.
- The increase in spectral, spatial and temporal resolutions including better calibration of data acquired by newly developed Earth orbiting sensors provides cost effective access to framework data and mapping bases.
- The introduction of personal computers and microelectronic equipment and telecommunication services (which gave access to Internet facilities) paving the way for an avalanche of information, not only for scientific research, but also for providing insight to a broader public, for planning or for policy purposes. The possibilities to query, retrieve, process and analyze remotely sensed data obtained via the Internet have aggrandized the interest of both remotely sensed data users and producers in many countries of Africa. The Internet has made it possible for end-users to more readily gather local geographic data germane to their own needs, draw data from library depositories elsewhere, develop the information products they need and use the data for decision making processes (WOLDAI, 2002).
- Besides remote sensing, innovations and advances achieved in Geoinformatics, GIS technology and the concepts of managing the Geo-information Infrastructure (GII) have provided tremendous advantage to various application fields in general and regional development planning in particular. GIS technology includes a range of tools and methods that not only preserve spatial context, but include modeling operations that make use of spatial configuration, such as proximity and adjacency. In this regard, GIS has offered a great capability for analyzing, integrating, modeling and presentations of spatial data, which in turn can be a good output for efficient and effective planning, evaluation as well as rational decision-making.

2 Infrastructure and available resources in support of RS development

2.1 Category of RS users

Thirty years after the technology was introduced in Africa, 3 categories of RS users are evident:

2.1.1 Category 1: The no-go situation

- To this scenario belong countries such as Angola, Burundi, Cameroon, Democratic Republic of Congo, Djibouti, Eritrea, Gambia, Liberia, People's Republic of Congo, Rwanda, and Somalia. These countries are poor, with some constantly experiencing drought. The countries have a history of civil wars; some sustained over long periods, with much consequent destitution, food insecurity and refugee problems. As such institutional capacity to promote RS technology is rather limited and at times absent.
- In general the existing infrastructure is rather poor and there are no resources available to support the remote sensing development.
- In nearly all, no identifiable RS center or surveying and mapping institute exist. NGO's and donor initiated RS activities often flourish and disappear when funding dries up.
- The geodetic network is far from complete.
- No baseline map (topographic, geological, vegetation, land use, etc) of the country exists other than uncontrolled photo-mosaics or reconnaissance sketches. If they do they are mostly outdated and at scale of 1:1 000 000 (minimal). The most basic maps are widely distributed over several government institutions that often lack coordination, seriously outdated with some going back to colonial periods.
- Skilled manpower is a major constraint.

Countries in the No-Go situation have to start at zero. The major question often asked by Governments and institutions in these countries is "where do we start?"

2.1.2 Category 2: The go situation

- The majority of the countries in Africa fall to this category.
- A Ministry, Agency, Authority or Center responsible for coordinating remote sensing activities is in place.
- Institutional weakness include excessive departmentalism, impractical allocation of power between local, regional and national government, the majority of the population being excluded from participation in decision-making and decision-making being largely unaccountable.
- There is considerable use of aerial photography, Landsat TM followed by SPOT imagery; quite often with no central archiving system but managed and owned by different organizations and Ministries who absolutely refrain from sharing.
- In many countries satellite interpretation is performed by RS specialists, often with little knowledge of the local conditions and in some cases even lacking the appropriate discipline background.
- Processing facilities might exist in the country but often protected with no cooperation whatsoever within organizations or institutions in the country.
- Research and development is poorly coordinated.
- Many foreign funded RS projects are not integrated within the operational procedures of the host government agency – with no demand for the information being supplied, the activities are not sustainable on project completion.
- In-country RS training courses and educational programs are done on an ad-hoc basis or at Centers of higher learning (e.g., University of Khartoum, University of Kenya etc.).

- In many cases, remote sensing hardware, such as computers, scanners, film-writers, printers, plotters, spectrometers, photo laboratory facilities, GPS etc. might exist. Although, the availability of hardware and software might differ from country to country, these days, due to technological advancement, it is possible to find in some institutions new versions of hardware and software almost every two year. Apparently, the number of software packages and hardware platforms in use even within a single department, let alone the whole country, causes operability problems in transferring data from one system to other. There is little or no provision for maintenance and running costs.

With the GO-GO situation, the need for an identifiable remote sensing center is high, but the will to implement the program is missing. The main question often asked is "where do we go from here?"

2.1.3 Category 3: The go-go situation

- To this category belong countries such as, South Africa, Morocco, Egypt and Zimbabwe etc.
- The Ministry responsible for coordinating Earth Observation (EO) program is well established.
- Policy, guidelines and decrees are well formulated. Research and development in remote sensing activities are well coordinated.
- It is pleasing to report that many of the technical and institutional constraints that stifle Earth observation today are already being bypassed by advances in micro-electronics. Both, the development need and the technology environments are becoming significantly more favorable for the transfer and local use of RS data.
- A country such as South Africa has an exceptionally successful and practical Earth observation program, managed largely by the users, achieved through a carefully planned multi-thrust approach initiated with targeted awareness rising and then aggressively pursued over a sustained period. The country can play an important role in the SADC region in information technology, provision of data, products and services and in training.
- South Africa has developed its own satellite and future possibilities for using this route for prompt dissemination of small satellite data sets are promising. Strategic investment in centralized reception, archive and subsidized distribution of RS data by the South African Satellite Application Center (SAC) has been a key factor in facilitating widespread access to and application of RS technology in the country.
- GIS capabilities exist. GIS using RS is becoming more and more important as the information extraction tool. Although within countries such as, Morocco, Egypt and Zimbabwe, there need to be some caution to ensure its efficient and effective use.
- Education and training is done at centers of higher education. Many Universities and Colleges in the GO-GO countries offer dedicated courses in RS and students are encouraged to pursue their MSc and PhD research using the technology.

2.2 Category of RS producers

With regards to categories 2 and 3, past experience has taught us that there are two types of remote sensing data producers in Africa: Those which produce the *foundation data*; basic spatial and non-spatial information and, those which produce *theme specific*

Geoinformation. In the first instance, the basic Geoinformation (GI) producers are those that are mandated at the national level for collecting, processing and disseminating GI and other related data at large. These include land administration authorities, surveying and mapping authorities, (municipal) planning departments, hydrological and topographic surveys; organizations traditionally involved in the production of Geoinformation data using remotely sensed data in the form of maps and reports. These organizations provide data that governments need in order to act. Their task is tied to the foundation data and as such deal with: the national positioning system, the national digital topographic template, digital elevation model, geographic names, administrative units, common codes, etc. It is an open secret that the national Geodetic/Topographic Surveying organization assuming monopolies are bestowed with the task of producing the baseline data that can serve most other GI productions. Administrative boundaries are often produced and maintained by the national statistical organization. The fact that such organizations are successful in satisfying the growing demand for simple and manageable remotely sensed base-information is however, questionable. Except for South Africa and to a certain extent countries like Côte d'Ivoiré and Ethiopia, surveying and mapping organizations have failed to supply digital GI products in a format, which most other organizations and individual users can easily use. The problem is manifold. On the one hand, in countries like Kenya, Sudan and to a certain extent, Botswana, Tanzania and Zambia, the baseline data can't satisfy the growing demand for digital base information despite having the legal rights and cover to monopolize the activity. On a second thought, most current products, mostly maps created using RS, are not even geo-referenced and as such not convenient for easy application. Moreover, the correctness of the existing topographic maps is questionable since updating the produced analogue outputs are difficult while some areas of their country are not yet covered totally. This really puts their usefulness in support for development decisions at stake.

The case with the cadastral offices or statistic administrations is not different, although disseminating the indicated products is not their prime objective. Surprisingly, if anyone wants to carry out tasks at a very lower administrative level, the only data one can get is in most cases the boundary information. Still the quality, especially the positional accuracy remains displeasing. What is more astonishing is that most such organizations in Africa haven't extended their services outside the capital cities and their environs. In fact, most of the organizations in Sub-Saharan Africa (excluding South Africa) have no branches at the regional levels. One can imagine how long it will take to cover a single region with only the national set available. Moreover, it is hardly possible for them to focus on RS products that can be supportive for regional development endeavors. It is not only the question of coverage and focus but if one wants to use the already available GI baseline data, one has to go as far as to the capital of the country where the main offices are located. In other words, this has to do with the fragmented chain of command and, most of all, with the poor accessibility to products in that one needs to reach them physically from whatever location one might be.

To the framework and theme specific Geo-data producers' category belong ministerial offices, authorities, commissions and project offices. Included in this category are civil society organizations such as community networks for disaster prevention, environmental control groups and local education and knowledge institutions. Most of these institutions use the baseline information from the various surveying and mapping organizations, hydrological authorities, meteorological institutions, statistical administrations, etc. to come up with their own value-added outputs. Depending on their thematic responsibility (e.g., geology, land use, land cover, vegetation etc.), they produce and use remotely sensed data on an ad-hoc basis (rather than taken as a regular duty) primarily to fulfill their jurisdictional activities. Unlike the GI basic users, most of these institutions, their bureaus and departments have a chance of getting more up-to-date data on their subject. Confined to their institutional mandates and concentrated on meeting their own internal information need, they have no mechanism to facilitate data access to other interested users. Therefore, unless it is a luckily informed user, no one can easily know what type of remotely sensed data they have, no matter how valuable the data are.

2.3 Policy and coordination

- **Need for RS information:** There is a clear need for timely information for decisions, but lack of awareness of these needs on the part of EO specialists, and lack of awareness of the potential of EO on the part of the decision makers, leaves a gulf that appears to be rarely bridged in spite of its obviousness. The comparative advantage of EO for time-sensitive decision making is largely lost in practice.
- **Duplication** of efforts, costly equipment and datasets emanating from stand-alone heterogeneous systems whereas the other side is suffering from dearth of information is still there.
- **Security of the RS data** holdings is still inadequate. All resources are stored in central location such that recovery in case of calamity would be difficult. Besides, an enormous amount of original RS documents has been destroyed due to poor storage facilities or remains in the shelves of various ministers or high officials unavailable to the wider user communities.
- **Minimal data exchange/sharing**. Improved analysis and decision making-as well as the cost-effective use of relatively expensive data-must spring from genuine cooperation and a willingness to share results. Although paper RS maps are usually available in many African countries, reluctance to share digital data and particularly digital images is common. To date, the only widely available digital maps for Africa are at 1:1 000 000. But analysts need a larger scale – that is, more detailed maps and data – to assess environmental conditions and trends and to propose solutions to environmental degradation at the national and sub-national levels.
- **Lack of meaningful co-ordination** among ministries, organizations and agencies prevails as the neglected 'Achilles heel' to the regional development planning effort of many Governments in Africa. In nearly all, Geoinformation activities derived from remotely sensed data are planned, implemented and monitored by different arms of Government (Ministry of Agriculture, Mining and Energy, Directorate of Surveying and Mapping, to name a few). They all specialize in various production aspects of surveying and mapping and have their own topographic & mapping facilities on piecemeal basis fragmented from the wider national issues. Institutional coordination in most is still absent. Many researchers and users of the RS technology suffer from isolation. Emerging RS capabilities also spawned uncontrollable new agencies, whose product have been poorly integrated with existing ones. The consequence of these is evident in the overlay of efforts, duplication of costly equipment and satellite data, which often induce competitions of scarce human and financial resources. Short-term

financial and macroeconomic issues took precedence over long-term renewable resources. So far, as money was coming from somewhere and the technology flourished otherwise, there was no long term commitments to the technology. Applications were customized to support specific decision-making process. As a result organizational focus remains by nature very narrow.

- **Rate of Change:** Environmental issues are becoming more and more pressing and rates of change are getting faster and faster. Monitoring rather than mapping is of the highest priority. Planning is becoming dynamic and strategic. Top-dawn overviews are required, with nested pockets of more detailed information. This minimizes the cost, is less sophisticated, and does not rely on only one information source, which may fail to provide the necessary results. Satellite imagery (both high and low resolution) and aerial photography, providing inputs to environmental information systems, will have a role to play in this model and the supply of near real-time information will become more critical. Apart from South Africa, which has an exceptionally successful and practical Earth Observation Program and who has developed its own satellite, Nigeria and Algeria have established space programs and launched their own mini-satellite with the technical help from the United Kingdom. Likewise, countries like Egypt, Morocco and Kenya are planning to install infrastructure for the reception of satellite imagery. Even in these countries that are advancing to the corporate stage, their efforts are entangled with various impediments, among those are the lack of technical manpower (adequate academic base) and a weak communication infrastructure.
- **Manpower Development:** Remote sensing and education is slowly been implemented in a number of colleges and universities in Africa. Most of these institutions however, are poorly equipped in regard to processing facilities and skilled manpower to provide in-depth education on RS. Investment in training people to use the technology, to collect, manipulate, and interpret geo-information is a 'must' but quite often receives lip-service. Years of investment on educators often result in experts leaving the university for a better pay somewhere. Retaining technical expertise should be a priority within African institutions. In many countries of Africa, non-academic organizations like national mapping agencies, geological companies, agricultural sector, and environmental authorities are more aware about environmental remote sensing than colleges or universities. Although some of these sectors train data users, its recognition is very limited. That is why one can hardly get coordinated approaches between non-educational and educational sectors to benefit from environmental remote sensing.
- **Communication Infrastructure:** The most daunting problem in Africa is the issue of communication. It is true, that the advancement in Information Systems/Information Technology (IS/IT), especially, the progress in the worldwide internet system has enabled the acquisition of a myriad of information that once was unthinkable to be acquired in Africa. Even then, the majority of the countries in Sub-Saharan Africa have only email access and cannot access the full facility of the internet. According to the PANOS Institute, and while the developed world is moving towards fibre optics, ISDN, and fifth generation computing, the developing world and in particular Africa still lacks functioning telephones and terminals. In a country such as Nigeria, which claims to have its own mini-satellite in orbit, it will be easy for one to get connected to Europe rather than within the country itself. Besides, on the continent where professional salaries are very low and where organizations put low priorities to

infrastructural set-up, modems are very costly and a personal computer represents many years of salary investment for the end user. Computers, if present in an organization are mostly bought by governments or large firms, which rely heavily on technical expertise, assistance, maintenance, and components from the First World, and thus it is for those reasons that PCs are often sitting underused or unused. Even in those countries, where the conditions for infrastructure are present, there is often a lack of programmers to create software for systems, thus forcing the reliance on software written in the First World. The dependence on the West for support with regard to computing, in terms of hardware, software, and training, has led many of African end-user to frustration. Rather than providing autonomy, it has become yet another aspect of underdevelopment and dependency.

3 Conclusion

The problems described above are part of the daunting situation facing most of the African countries. A well-structured infrastructure and resources in supporting RS development are still missing. The picture sketched above regarding RS technology in Africa must be taken as a lesson if efforts towards the effective usage of the technology are to succeed. Having said that, the implementation of a proper RS infrastructure and resources in Africa requires the political will at the highest echelon of Governments; this should comprise a solid infrastructure based on policy, guidelines and administrative arrangements, and effective means by which remotely sensed spatial data are made accessible to the community. The past thirty years of RS technology in Africa should be the lesson and a proof that the initiative is doomed to fail without such commitment.

References

PANOS Institute, "The Internet and the South", online document. http://www.oneworld.org/panos/internet-and-south.html.

WOLDAI, T., 1995, Survey and mapping in Sub-Saharan Africa – Which way further? Invited paper, UN-Outer Space Affairs Division (OSAD), United Nations Regional Conference, Senegal, West Africa.

WOLDAI, T., 2002a, Invited Paper: The Status of Environmental Information Systems in Sub-Saharan Africa, Developments and Technology Transfer in Geomatics for Environmental and Resource Management, ISPRS Commission VI Workshop, Dar-es Selaam, Tanzania.

WOLDAI, T. and SCHETSELAAR, E., 2002b, Design And Implementation Of "Factual" Databases to Support GIS/RS Analyses of Earth Systems For Civil Society, Developments and Technology Transfer in Geomatics for Environmental and Resource Management. ISPRS Commission VI Workshop, Dar-es Selaam, Tanzania, pp. 117–124.

Chapter 5

Summary of Final Panel Discussion

Referring to global frameworks such as the action plan of the World Summit on Sustainable Development, the Millennium Development Goals, *the key objectives of development are peace, prosperity and dignity* (A. Wright). The initiative, which led to the organisation of this event (gi4dev), is seen as best practice for exchange and reflection on questions such as:

Are we prepared to learn the lessons?

And, even having identified the right things to act on, are we doing these things right?

For geoinformation to have an impact on development, the panel and the audience stated that the following factors are crucial:

- Sustainability must be considered right from the onset of a project and the sense of ownership is vital for success. As brain drain is a problem, the limited capacities in the developing world need to be strengthened to exercise ownership, in expert organisations as well as communities.
- Do we have a common understanding what 'assistance' means and are we actually causing an impact on development? Besides the coordination between donors as well as developers and implementers, a key requirement is that partners formulate research (or development) questions together. Balanced partnerships should be thematic, political, research oriented, and need to bridge natural and social sciences to ascertain the societal relevance of geoinformation systems.
- Project partners have to be clear about stakeholders, beneficiaries and key benefits to assure development for all. For instance, geoinformation infrastructure effectively supports decision-making for risk reduction only when reducing the vulnerability of all societal groups (J.C. Villagrán de León, D. Sakulski).
- A strategic approach is required to integrate data, knowledge and culture. Transparent and systemic planning facilitates the achievement of results, which may not always represent the optimal solution, but have to be practical, clear and simple.
- Development and implementation of Geoinformation Systems must include people, technology and policies.

The participants concluded that future gi4dev conferences should monitor and evaluate the implementation of these requirements.

The structure of the conference

To facilitate the exchange of experience and the flow of discussions as best as possible, we selected the following format for the three thematic sessions:

A keynote at the beginning of each session introduced the topic based on a case study (15 min + 5 min discussion). The speakers of the following papers gave an overview about the main issues (5 min each) to the plenary followed by an in-depth discussion aided by laptop and posters (30 min) at four individual stands distributed in the room. At the end of each session, the main points were summarised with the support of the session's moderator, who also compiled action points for further follow-up.

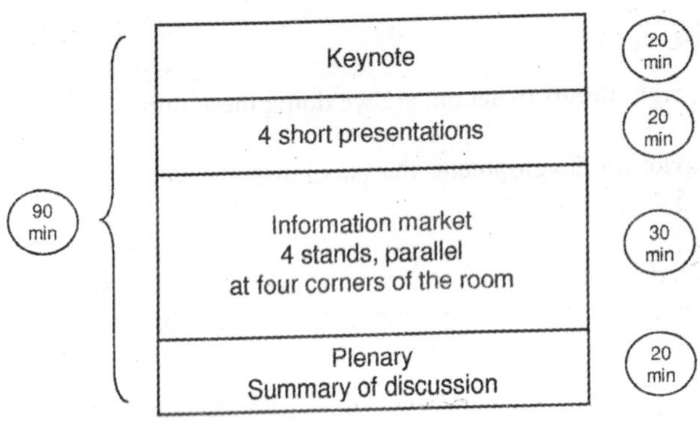

Fig. 1: Session structure for gi4dev 2006

he plenary

Fig. 3: In depth discussion

Fig. 4: Final panel discussion

Fig. 5: Panel members (from left: F. Hörlberger, J. C. Villagrán de León, E. A. Wright, D. Stevens)

List of Authors and Contact Details

Karl Atzmanstorfer
University of Salzburg
Hellbrunnerstrasse 34
5020 Salzburg, Austria
karl.atzmanstorfer@sbg.ac.at
MSc-student; GIS-application to regional development issues in developing countries (Latin America), international development cooperation

Alexander Bohr
IDC/ European Commission
Avenida Frederich Engels 363/1
Maputo, Mozambique
alexander.bohr@odamoz.org.mz
Coordination and harmonisation of the ODAmoz database capturing financial information related to ODA flows to Mozambique

Federico Breda
Istituto Agronomico per l'Oltremare IAO
Via Cocchi 14
Florence, Italy
breda@iao.florence.it, federico.breda@dappolonia.it
GIS/RS analyst

Sebastien Delhay
Signal and Image Centre (SIC)
Royal Military Academy
Avenue de la renaissance 30
1000 Bruxelles, Belgium
sdelhay@elec.rma.ac.be
Sebastien Delhay is an industrial engineer. At the SIC he has developed a GIS interface for the management of humanitarian demining campaigns.

Robert Etim Ekpenyong
University of Uyo
Department of Geography and Regional Planning
Uyo-Akwa Ibom State, Nigeria
robert_etim@yahoo.com
Senior Lecturer in the field of Geography with specialization in cartography and Geographic Information Systems (GIS)

Ernst Gebetsroither
ARC systems research GmbH
Donau-City-Strasse 1
Vienna, Austria
Ernst.Gebetsroither@arcs.ac.at
Environmental planning

Carsten Haub
EFTAS Fernerkundung Technologietransfer GmbH
Ostmarkstrasse 92
Münster, Germany
carsten.haub@eftas.com
Carsten Haub is leading the EFTAS section Landscape Ecology and Land Resource Management (www.eftas.com) and is GMFS regional task manager for East Africa (www.gmfs.info)

Ges. Mag. Franz Hörlberger
Former Head of Department VII.3
Humanitarian Assistance, NGO-Co-financing and Business Development
Austrian Ministry for Foreign Affairs
Now: Ambassador of the Republic of Austria, Hashemite Kingdom of Jordan
Mithqal Al-Fayez Street 36, Jabal Amman
P.O.B. 830795, Amman 11183, Jordan

Stefan Kienberger
Centre for Geoinformatics
Salzburg University
Hellbrunnerstrasse 34
5020 Salzburg, Austria
stefan.kienberger@sbg.ac.at
Disaster Risk Management, Vulnerability Assessment, Participatory GIS, GIS and developing countries

Dr. Diethard Leber
Department of Environmental Geosciences
Center for Earth Sciences
University of Vienna
Althanstrasse 14
1009 Vienna, Austria
diethard.leber@univie.ac.at
Integrated analysis of natural hazards and mitigation strategies, Groundwater systems and water supply, Remote Sensing Application

Dr. Wolfgang Loibl
ARC systems research
Donaucitystraße 1
1220 Vienna, Austria
wolfgang.loibl@arcs.ac.at
Main interests are modelling spatial phenomena – currently focussed on simulation of effects of human – environment interaction, on spatial effects of human intervention into environmental systems and on regional climate modelling and climate change impact research

Ralf Löwner
GeoForschungsZentrum Potsdam
Department 4.2 and Data Center
Telegrafenberg A3
14473 Potsdam, Germany
loewner@gfz-potsdam.de
Geologist and specialist in data and information management, especially in spatial data infrastructures for science, coordinator of the IT working group and of the German part of the SaDIN project

Dr. Jan-Peter Mund
Senior Lecturer at the Royal University of Agriculture, Phnom Penh, Cambodia
c/o GTZ Office Phnom Penh, P. O. Box 81
Phnom Penh, Cambodia
jpmun03@yahoo.com
Dr. Mund is a physical geographer and senior lecturer with specialisation in landscape ecology and sustainable ecological development issues as well as applied geographical information systems and remote sensing applications in Africa and South-East Asia

Norbert Niederhauser
Centro Internacional de Agricultura Tropical – CIAT
A.A. 6713, Km 17 Recta Cali – Palmira
Cali, Colombia
n.niederhauser@cgiar.org
Data management and information engineering for development projects

Dr. Rachel O'Brien
Charles Sturt University
School of Environmental and Information Sciences, Charles Sturt University, PO Box 789, Albury NSW 2640, Australia
robrien@csu.edu.au
GIS modelling, species distribution modelling, spatial decision support systems, uncertainty and risk in GIS, spatial algorithms and programming

Richard Resl
Universidad San Francisco de Quito
Redondel Cumbaya, Ecuador
Casilla 17 – 22 – 20205
Quito, Ecuador
unigis.admin@mail.usfq.edu.ec
Geographer, GIS consultant and Coordinator of UNIGIS distance learning program in GIS, GISciudadano and AmazonGISnet in Latin America

Gernot Rücker
ZEBRIS GbR
Lipowskystr. 26
81373 München, Germany
gruecker@zebris.com
Managing director (www.zebris.com)

Dr. Dusan Sakulski
United Nations University – Institute for Environment and Human Security
Hermann-Ehlers-Str. 10
53113 Bonn, Germany
sakulski@ehs.unu.edu
Academic Officer Innovative Technologies for Disaster Risk Reduction

Prof. Dr.-Ing. Gertrud Schaab
Hochschule Karlsruhe – Technik und Wirtschaft, Fakultät für Geomatik
(Karlsruhe University of Applied Science, Faculty of Geomatics)
Moltkestraße 30
76133 Karlsruhe, Germany
gertrud.schaab@hs-karlsruhe.de
Gertrud Schaab is especially interested in the visualisation and simulation of spatial processes and more broadly in interdisciplinary applications of GI-technology in environmental sciences

Harald Schernthanner
University of Potsdam/ Institute of Geography/ GIS department
Jungstraße 12
10247 Berlin, Germany
hschernt@uni-potsdam.de
Teaching of remote sensing and GIS at the University of Potsdam/Specialisation in applied geographic information services and remote sensing

Dr. Wilko Schweers
Chinese Academy of Agricultural Sciences – CAAS
Institute of Agricultural Resources and Regional Planning
No. 12 Zhongguancun South Avenue
100081 Beijing, China
w_schweers@caas.ac.cn
Sustainable Management of Scarce Groundwater Resources in the Agriculture of Northern China

David Stevens
United Nations Office for Outer Space Affairs (UNOOSA)
Vienna, Austria
david.stevens@unvienna.org
David Stevens is a Programme Officer of UNOOSA where he is responsible for promoting the use of space technology for disaster management and complex emergencies in developing countries. His area of expertise is the use of geo-spatial information solutions to support development activities. His home country is Brazil.

Dr.-Ing. Harry Storch
Brandenburg Technical University (BTU) Cottbus
Faculty 4, Environmental Sciences and Process Engineering
Department of Environmental Planning
Erich-Weinert-Str. 1
03044 Cottbus, Germany
storch@tu-cottbus.de
Dr. Harry Storch is working as research assistant in the project 'Sustainable Housing Policies for Megacities of Tomorrow – The Balance of Urban Growth and Redevelopment in Ho Chi Minh City' which is financed as part of the new research programme 'Megacities of Tomorrow' by the German Federal Ministry of Education and Research (BMBF)

Ferdinando Urbano
Via F.Lli Pozzi 9
20100 Milano, Italy
feurbano@tin.it
Consultant

Juan Carlos Villagrán de León
Director: Centro de Investigación y Mitigación de Desastres Naturales CIMDEN-VILLATEK
15 Avenidad "A" 20-01 Zona 13
Guatemala City, Guatemala
jcv@villatek.com.gt
Dr. Juan Carlos Villagrán de León holds an M.S and Ph.D. degrees from the University of Texas at Austin in the field of experimental condensed matter physics. He has been actively involved on issues of risk assessment, risk management, and early warning in Guatemala and throughout Central America during the last decade, and is the author of several publications on these subjects.

Dr. Tsehaie Woldai
President, African Association of Remote Sensing of the Environment (AARSE)
Associate Professor Geological Remote Sensing, Portfolio Manager Education
Department of Earth Systems Analysis
International Institute for Geo-Information Science and Earth Observation (ITC)
Hengelosestraat 99
PO Box 6
7500 AA Enschede, The Netherlands
Woldai@itc.nl

Prof. Dr.-Ing. E. Alaphia Wright
UNESCO/University of Zimbabwe
8 Rue Paul Hervieu
75015 Paris, France
abwright@noos.fr
Former Dean of Engineering, University of Zimbabwe. Currently responsible for Evaluation Capacity Development, UNESCO

Peter Zeil
Centre for Geoinformatics
Salzburg University
Hellbrunnerstrasse 34
5020 Salzburg, Austria
peter.zeil@sbg.ac.at
Geophysicist with 25 years of professional experience in Europe, Africa and Latin America related to geoinformation and remote sensing. Presently, he is a senior project manager at the Centre for Geoinformatics (Z_GIS), Salzburg University (Austria) and coordinates several international research projects.

UNITED NATIONS UNIVERSITY
Institute for Environment and Human Security (UNU-EHS)

UNITED NATIONS UNIVERSITY

UNU-EHS

Institute for Environment and Human Security

UNU-EHS reflects the overall mission of UNU: 'Advancing Knowledge for Human Security and Development'. UNU-EHS explores threats to human security from environmental degradation, unsustainable land use practices, and from natural and human-induced hazards. The Institute spearheads UNU's research and capacity building activities in the broad interdisciplinary field of 'risk and vulnerability'.

Within this framework UNU-EHS aims to:

- Foster better understanding of forces and processes of environmental degradations and their influence on hazard magnitude and frequency and subsequent disasters;
- Explore links between different hazard events as well as creeping processes such as climate change, soil erosion and their impact on the inherent risk and vulnerability;
- Contribute to development, testing and verification of vulnerability indicators, and investigate relationships between risks, vulnerability and coping capacity;
- Investigate the relationship between risk, vulnerability and coping capacity, and devise strategies and measures to improve the coping capacity of affected communities.

Its activities focus on:

- Generating knowledge;
- Building capacity, with focus on young professionals and institutions in developing countries;
- Disseminating knowledge to decision makers.

For further information: http://www.ehs.unu.edu/

Chapter 5

Summary of Final Panel Discussion

Referring to global frameworks such as the action plan of the World Summit on Sustainable Development, the Millennium Development Goals, *the key objectives of development are peace, prosperity and dignity* (A. Wright). The initiative, which led to the organisation of this event (gi4dev), is seen as best practice for exchange and reflection on questions such as:

> Are we prepared to learn the lessons?
>
> And, even having identified the right things to act on, are we doing these things right?

For geoinformation to have an impact on development, the panel and the audience stated that the following factors are crucial:

- Sustainability must be considered right from the onset of a project and the sense of ownership is vital for success. As brain drain is a problem, the limited capacities in the developing world need to be strengthened to exercise ownership, in expert organisations as well as communities.
- Do we have a common understanding what 'assistance' means and are we actually causing an impact on development? Besides the coordination between donors as well as developers and implementers, a key requirement is that partners formulate research (or development) questions together. Balanced partnerships should be thematic, political, research oriented, and need to bridge natural and social sciences to ascertain the societal relevance of geoinformation systems.
- Project partners have to be clear about stakeholders, beneficiaries and key benefits to assure development for all. For instance, geoinformation infrastructure effectively supports decision-making for risk reduction only when reducing the vulnerability of all societal groups (J.C. Villagrán de León, D. Sakulski).
- A strategic approach is required to integrate data, knowledge and culture. Transparent and systemic planning facilitates the achievement of results, which may not always represent the optimal solution, but have to be practical, clear and simple.
- Development and implementation of Geoinformation Systems must include people, technology and policies.

The participants concluded that future gi4dev conferences should monitor and evaluate the implementation of these requirements.

The structure of the conference

To facilitate the exchange of experience and the flow of discussions as best as possible, we selected the following format for the three thematic sessions:

A keynote at the beginning of each session introduced the topic based on a case study (15 min + 5 min discussion). The speakers of the following papers gave an overview about the main issues (5 min each) to the plenary followed by an in-depth discussion aided by laptop and posters (30 min) at four individual stands distributed in the room. At the end of each session, the main points were summarised with the support of the session's moderator, who also compiled action points for further follow-up.

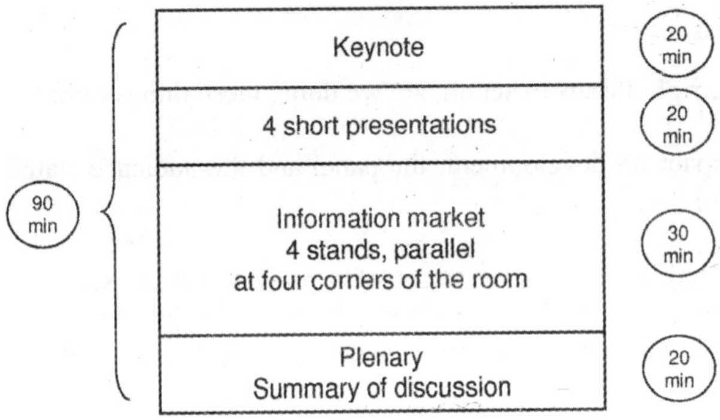

Fig. 1: Session structure for gi4dev 2006

Fig. 2: The plenary